SpringerBriefs in Applied Sciences and Technology

SpringerBriefs present concise summaries of cutting-edge research and practical applications across a wide spectrum of fields. Featuring compact volumes of 50 to 125 pages, the series covers a range of content from professional to academic.

Typical publications can be:

- A timely report of state-of-the art methods
- An introduction to or a manual for the application of mathematical or computer techniques
- A bridge between new research results, as published in journal articles
- A snapshot of a hot or emerging topic
- An in-depth case study
- A presentation of core concepts that students must understand in order to make independent contributions

SpringerBriefs are characterized by fast, global electronic dissemination, standard publishing contracts, standardized manuscript preparation and formatting guidelines, and expedited production schedules.

On the one hand, **SpringerBriefs in Applied Sciences and Technology** are devoted to the publication of fundamentals and applications within the different classical engineering disciplines as well as in interdisciplinary fields that recently emerged between these areas. On the other hand, as the boundary separating fundamental research and applied technology is more and more dissolving, this series is particularly open to trans-disciplinary topics between fundamental science and engineering.

Indexed by EI-Compendex, SCOPUS and Springerlink.

Hairus Abdullah

Vanadium Oxide-Based Cathode for Supercapacitor Applications

Using Electrodeposition Method

 Springer

Hairus Abdullah
Department of Materials Science
and Engineering
National Taiwan University of Science
and Technology
Taipei, Taiwan

Department of Industrial Engineering
Faculty of Science and Technology
Universitas Prima Indonesia
Medan, Indonesia

ISSN 2191-530X   ISSN 2191-5318   (electronic)
SpringerBriefs in Applied Sciences and Technology
ISBN 978-981-97-5242-3   ISBN 978-981-97-5243-0   (eBook)
https://doi.org/10.1007/978-981-97-5243-0

This Springer imprint is published by the registered company Springer Nature Singapore Pte Ltd.
The registered company address is: 152 Beach Road, #21-01/04 Gateway East, Singapore 189721,
Singapore

If disposing of this product, please recycle the paper.

Preface

Energy crisis and environmental issues are the most crucial topics in recent years. Non-renewable fossil fuels not only bring ecological problems, but also impact the global economy and society. Implementing solar cell technology has been the solution to overcome environmental issues. However, the intermittent of sunlight poses a significant challenge in harnessing the solar light energy. Technological development of energy storage materials in supercapacitors and batteries is one of the crucial tasks to support the use of green energy sources. An immediate need exists for energy storage devices that are not only efficient but also environmentally sustainable. These devices should be capable of supplying power to energy-intensive sectors, including transportation and portable gadgets, for example, laptops, camcorders, mobile phones, electric vehicles, and hybrid electric vehicles. The devices such as fuel cells, solar cells, photoelectrochemical water splitting cells, batteries (particularly Li-ion batteries), and supercapacitors are typical energy storage technologies. The functionality of these energy devices is substantially influenced by the characteristics of nanostructured materials. It is hypothesized that advancements in nanomaterial chemistry may provide the means to achieve additional breakthroughs in energy storage systems. Owing to the large surface-to-volume ratios, favorable transport properties, altered physical properties, and confinement effects resulting from their nanoscale dimensions. Therefore, nanostructured materials have been the subject of extensive research. Vanadium-based electrode emerges as a promising energy material. The vanadium electron configuration is $3d^3\,4s^2$ and all five valence electrons in bonding processes result in the formation of multivalent V, ranging from V^{2+} to V^{5+}. This multivalent V contributes to the formation of vanadium oxides, vanadates, and vanadium phosphates. In addition to the abundant source and rich electrochemical characteristics of V (V^{2+} to V^{5+}), V-based nanomaterials are also economically viable and a promising electrode candidate in the field of energy storage.

In this book, the related backgrounds of supercapacitors and the basic indicators to examine supercapacitor performances are briefly discussed in Chap. 1. Further state of art works in developing supercapacitors is provided in Chap. 2. Some fundamental regulations for improving the electrochemical properties are also discussed in the chapter. An example of a typical work of vanadium oxide-based cathode for a supercapacitor prepared by an electrodeposition method is provided in Chap. 3. A further advanced improvement of electrodeposited vanadium oxide supercapacitor with bilayer $Ni(OH)_2/VO_x$ is demonstrated in Chap. 4. In addition, Chap. 5 provides the advances of vanadium oxide supercapacitor works are discussed by providing different phases of VO_x with various morphologies to extend and develop the V-based supercapacitor works. Finally, the author acknowledges Program Lembaga Penelitian/Pengabdian Masyarakat (LPPM), Universitas Prima Indonesia with grant number #43/PL/2024 dated 23rd January 2024 in supporting the present work.

Taipei, Taiwan   Hairus Abdullah

Contents

Abbreviations

AC	Active carbon
CNT	Carbon nanotube
CPE	Constant-phase element
Cs	Specific capacitance
CV	Cyclic voltammogram
DRS	Diffuse reflectance spectroscopy
EDLC	Electrostatic double-layer capacitor
EIS	Electrochemical impedance spectroscopy
FE-SEM	Field emission-scanning electron microscopy
GCD	Galvanostatic charge-discharge
HCP	Hexagonal closed-packed
HER	Hydrogen evolution reaction
K_B	Boltzmann constant
LED	Light emitting diode
MO	Metal oxide
MWCNT	Multi-walled carbon nanotube
PANI	Polyaniline
PEDOT	Poly(3,4-ethylenedioxythiophene)
PPy	Polypyrrole
PVA	Polyvinyl alcohol
Rct	Charge transfer resistance
Rs	Internal series resistance
Rt	Tandem resistance
SAED	Selected area electron diffraction
SC	Supercapacitor
TEM	Transmission electron microscopy
TMO	Transition metal oxide
VN	Ni-doped vanadium cathode
XPS	X-Ray photoelectron spectroscopy
XRD	X-ray diffraction

Chapter 1
Introduction

1.1 Backgrounds

The advancement of science and technology in recent years with 5G technology has indeed affected the development of society. However, highly efficient technology has increased the demand for energy consumption by pursuing more functions in applications. Therefore, energy supply components on devices are receiving more and more attention. Consistently, developing energy storage components becomes crucial to support advanced technology. In addition, energy storage components should have high energy density, fast charge and discharge, and long lifetime [1]. In order to comply with the above requirements, the development of supercapacitors was initiated. Supercapacitors are widely used in energy supply, vehicle transportation, mobile communications, marine transportation, national defense, and military fields. Generally, energy storage components are mainly batteries and traditional capacitors. Batteries have high energy density, but traditional capacitors have high power density, and supercapacitors have the characteristics of both traditional capacitors and batteries. Supercapacitor (SC), also known as electrochemical capacitor [2], has a higher capacitance than traditional capacitors due to its special energy storage mechanism. In addition, as the charge accumulates on the surface of electrodes, SC has a higher power density and faster charge and discharge process compared to batteries. SC has not only reversible characteristics similar to capacitors, but also a longer cycle life than batteries. Traditional capacitors have the advantages of high power density, fast charge and discharge, and long lifetime, but they have the disadvantages of low energy density. On the other hand, batteries have high energy density but low power density, long charge and discharge time, and poor cycle life. Supercapacitors have the advantages of high energy density, high power density, fast charge and discharge, and long cycle life. A specific Ragone plot can be seen in Fig. 1.1 [3].

Although supercapacitors have many advantages, they are still in the research stage. At this stage, most use the solvothermal synthesis method to fabricate active materials [4]. However, this method is unsuitable for mass manufacture, making its

H. Abdullah, *Vanadium Oxide-Based Cathode for Supercapacitor Applications*, SpringerBriefs in Applied Sciences and Technology, https://doi.org/10.1007/978-981-97-5243-0_1

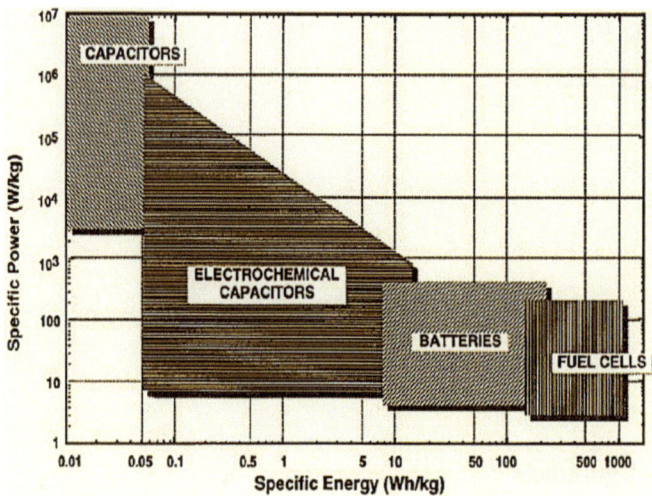

Fig. 1.1 Ragone diagram of energy and power densities of different energy storage devices [3] (Springer Nature Copyright 2012 with license number: 5772221189317)

industrialization difficult. Therefore, how to solve the manufacturing process issues still needs further discussion. In this chapter, the electroplating method was used to deposit a V_3O_5 thin film on the current collector. The manufacturing process is simpler than the solvothermal synthesis method and suitable for mass manufacturing. At present, the electroplating method is quite mature in terms of industrial process technology. If it can be developed successfully, the industrial development of supercapacitors will have significant progress in the future.

1.2 Classification of Supercapacitor and Mechanism

Based on the mechanism pathways, the supercapacitor can be classified as electrostatic double-layer capacitors (EDLC) and pseudocapacitors [5, 6].

1.2.1 Electrostatic Double-Layer Capacitors (EDLC)

When a voltage is applied to the two electrodes, the anions and cations in the electrolyte will be adsorbed on the surface of the positive and negative electrodes in a relative arrangement, as shown in Fig. 1.2, forming a double charge layer phenomenon. When the voltage applied to the two electrodes is removed, following a connection of the two electrodes with loading, the double charge layer ions previously accumulated on the electrode will move into the solution, causing charge neutralization, changing

Fig. 1.2 Electrostatic
double-layer capacitor
(EDLC) behavior when a
potential is applied to the
electrodes

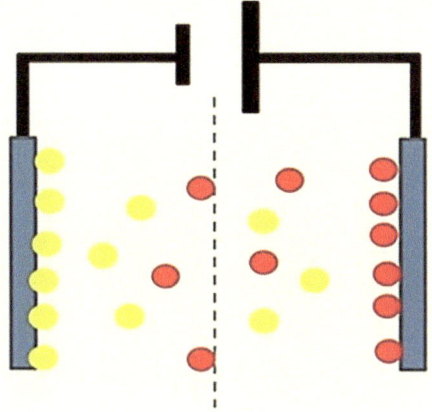

the potential, and releasing energy. As the mechanism of the electrostatic double-layer is basically due to the adsorbed ions on electrode surfaces, its capacitance is related to the use of the electrolyte. Electrode materials are closely related to and mainly affected by specific surface area and porosity. Therefore, carbon materials with high specific surface area or high porosity are often used as electrodes in this type of capacitor.

1.2.2   Pseudocapacitors

In addition to storing energy with a small amount of physically adsorbed electrical double-layer charge on the electrode surface, pseudocapacitors mainly utilize continuous and rapid redox electrochemistry between the active materials covered on the electrode surface and the electrolyte. The redox reaction involves a particular number of electrons, forming a stored energy. As it involves the Faradaic charge transfer reaction on the overall active material, the amount of charge stored is much greater than that of traditional capacitors or electrostatic double-layer capacitors. Transition metals in various valence states can produce a large number of Faradaic charge transfer reactions with electrolytes, they are often used as active materials, such as manganese oxide, nickel oxide, and vanadium oxide, as shown in Fig. 1.3. Among them, vanadium oxide has attracted the attention of most scientists. Among most transition metals, vanadium (V) has considerable potential because V has multiple oxidation states, such as 2+, 3+, 4+, and 5+, with the propitious characteristic of cheap, simple, and low in toxicity.

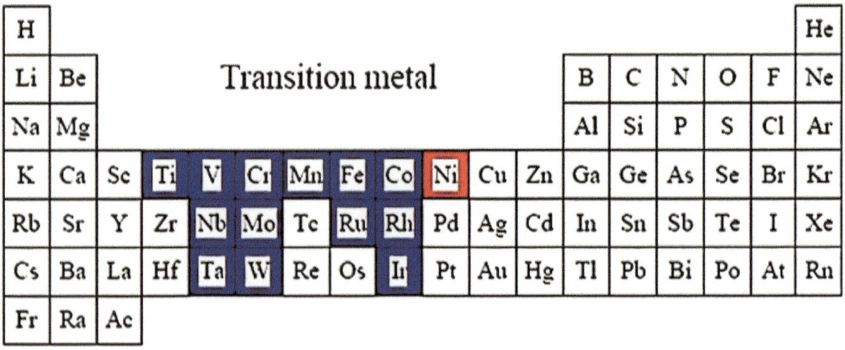

H																	He
Li	Be			Transition metal								B	C	N	O	F	Ne
Na	Mg											Al	Si	P	S	Cl	Ar
K	Ca	Sc	Ti	V	Cr	Mn	Fe	Co	Ni	Cu	Zn	Ga	Ge	As	Se	Br	Kr
Rb	Sr	Y	Zr	Nb	Mo	Tc	Ru	Rh	Pd	Ag	Cd	In	Sn	Sb	Te	I	Xe
Cs	Ba	La	Hf	Ta	W	Re	Os	Ir	Pt	Au	Hg	Tl	Pb	Bi	Po	At	Rn
Fr	Ra	Ac															

Fig. 1.3 Transition metals commonly used in supercapacitors

1.3 Indicators of Supercapacitor Performances

1.3.1 Definition of Capacitor

The unit of capacitance (Capacitance, C) is Farad (F), which is defined as unit voltage (V)/the charge (Coulomb, Q), as shown in Eq. (1.1).

$$C = \frac{Q}{V} \tag{1.1}$$

The specific capacitance of materials can be estimated based on cyclic voltammetric (CV), Galvanostatic charge–discharge (GCD), and electrochemical impedance spectroscopy (EIS).

1.3.2 Cyclic Voltammetry (CV)

Cyclic voltammetry is one of the powerful tools for measuring the redox potential of materials. Its principle is to scan at a fixed scan rate (V/s) from a starting potential to a positive potential or a negative potential in a selected potential range. When the material undergoes a redox reaction at a certain potential, the internal charge is converted, so the current I and voltage V will break away from the linear relationship and generate a response current, as seen in Eq. (1.2) with a given redox potential. As the energy storage mechanisms are different, the corresponding diagrams are also different. Figure 1.4a is the CV curve of an ideal electric double-layer capacitor because the energy storage mechanism of the electric double-layer capacitor is purely dominated by physical adsorption, without internal charge conversion; therefore, the graph will be approximately rectangular. Figure 1.4b is the CV curve of an ideal pseudocapacitor capacitor. The energy storage method is an oxidation/reduction

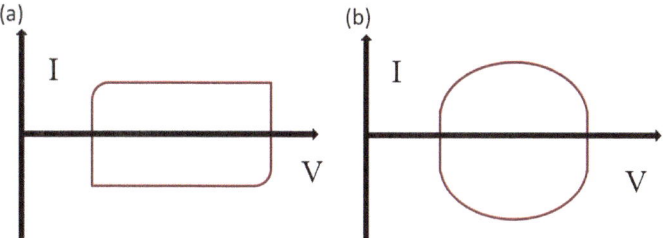

Fig. 1.4 Characteristic cyclic voltammogram of **a** EDLC and **b** pseudocapacitor

reaction between the electrode and the electrolyte. When a response current changes, the graph will deviate from the rectangle shape. The specific capacitance value can be calculated from the integrated area under the *CV* curve, as shown in Eq. (1.3).

$$i(t) = CV\left(1 - e^{-t/R_c}\right) \tag{1.2}$$

in which C is specific capacitance (F), V is the potential (V), R_c is contact resistance (Ω), and t is time (s).

$$C = \frac{\int I(V)dV}{(m \cdot v \cdot \Delta V)}(F/g) \tag{1.3}$$

in which I is the current (A), v is the scan rate (mV/s), m is the loading mass (g), and ΔV is the potential window (V)

1.3.3 Galvanostatic Charge/Discharge (GCD)

Galvanostatic charge/discharge is mainly used to evaluate the amount of charge stored inside the material. The principle is to apply a fixed current, the electrode will absorb and store the charge, causing the potential to rise. A corresponding current will be applied when the potential rises to the preset potential. The charge is released, causing the potential to drop. Depending on the energy storage mechanism, the graphic display is also different. Figure 1.5a shows the ideal electrostatic double-layer capacitor because the electrode is dominated by physical adsorption, and the slope of the voltage and time curve is constant, which is a linear relationship. Figure 1.5b indicates a galvanostatic charge/discharge diagram of an ideal pseudocapacitor capacitor. As the mechanism involves the conversion of charges with redox reactions, the slope of the voltage and time curve is not constant (nonlinear relationship). In addition, there is a so-called internal resistance in the capacitor. This internal resistance comprises the electrolyte resistance, the contact resistance between the electrode and electrolyte, and charge transfer resistance. This internal resistance will cause a certain resistance during the initial discharge. As a result,

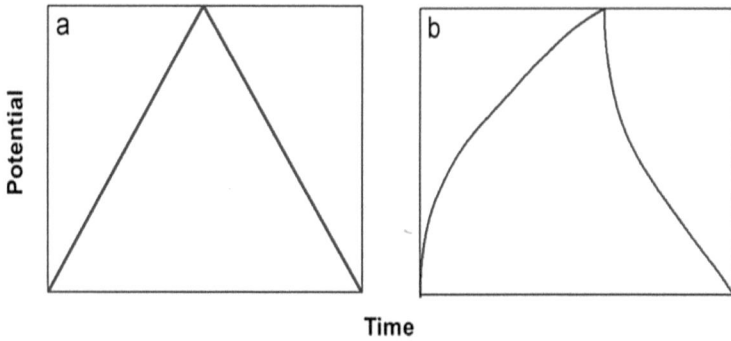

Fig. 1.5 Galvanostatic charge–discharge (GCD) diagram of **a** EDLC and **b** pseudocapacitor due to quasi-capacitance effect

in the galvanostatic charge/discharge diagram, there will be a potential drop at the starting position of the discharge, which is called an IR-drop. In addition, we can use the discharge time obtained from the galvanostatic charge/discharge diagram to calculate the capacitance value, as shown in Eq. (1.4).

$$C = \frac{I\Delta t}{\Delta Vm}(\text{F/g}) \tag{1.4}$$

in which I is the current (A), $\Delta V/\Delta t$ is the scan rate (mV/s), and m is the loading mass (g)

1.3.4 Alternating Current (AC) Impedance

AC impedance is also called electrochemical impedance spectrum, which can be used to analyze the electrode interface, structure, and reactions of the electrochemical system. As shown in Fig. 1.6, when a system is subjected to interference, it will produce a corresponding response, represented by $Y = GX$, where Y is the response signal, G is the transfer function, and X is the disturbance signal. As the system is an electrochemical reaction system, it can be expressed by $E(\omega) = Z(\omega)I(\omega)$. When an AC voltage $E(\omega)$ is given with a corresponding AC current $I(\omega)$, the AC impedance $Z(\omega)$ of the system can be obtained through Ohm's law, which is a sine wave function related to the frequency ω.

The electrode interface with a Faraday charge transfer resistance at low AC frequencies will show a straight line with a fixed slope in the impedance curve, suggesting the resistance of the electrochemical system. In the schematic diagram of Fig. 1.6a, R_Ω is the resistance of the electrolyte, C_d is the capacitance caused by the physical adsorption of charges at the electrode interface, and Z_f is the impedance caused by the AC charge transfer between the electrolyte and electrode. Figure 1.6b

(a)   (b)

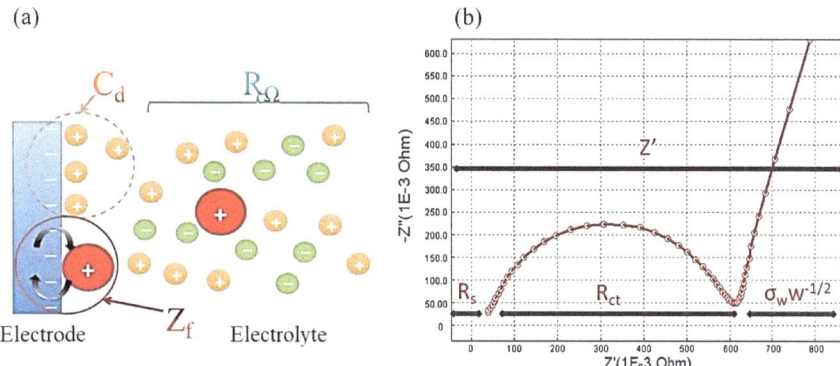

Fig. 1.6 a Physical adsorption of ionic species on electrode interfaces and **b** the simulated EIS curve to indicate the series (Rs) and charge transfer (R_{ct}) resistances

shows the electrochemical impedance spectroscopy (EIS) curve with the first point at abscissa (Re(Z) axis) at a high-frequency region is the ion transfer resistance in the electrolyte, also called internal resistance (Rs). While in the high-frequency region, the semicircle curve is the charge transfer resistance (R_{ct}) at the interface. As the frequency decreases, the straight line after the semi-arc curve is the mass transfer resistance of ions (equivalent distributed pore resistance, $\sigma_w w^{-1/2}$). Based on the equivalent circuit simulation, the equivalent series resistance (tandem resistance, R_t) in Fig. 1.6b can be obtained by complying with $Z\prime = R_s + R_{ct} + \sigma_w w^{-1/2}$. $Z\prime$ is essential to the capacitor behavior, indicating the iR drop in the GCD curve.

References

1. Q. Zhang, X. Xu, S. Chen, G.B. Bodedla, M. Sun, Q. Hu, Q. Peng, B. Huang, H. Ke, F. Liu, T.P. Russell, X. Zhu, Phenylene-bridged perylenediimide-porphyrin acceptors for non-fullerene organic solar cells. Sustain. Energy Fuels **2**(12), 2616–2624 (2018)
2. P. Li, Y. Yang, E. Shi, Q. Shen, Y. Shang, S. Wu, J. Wei, K. Wang, H. Zhu, Q. Yuan, A. Cao, D. Wu, Core-double-shell, carbon nanotube@polypyrrole@MnO$_2$ sponge as freestanding, compressible supercapacitor electrode. ACS Appl. Mater. Interfaces **6**(7), 5228–5234 (2014)
3. J. Kunze-Liebhäuser, O. Paschos, S.S. Pethaiah, U. Stimming, Fuel cell comparison to alternate technologies, in *Encyclopedia of Sustainability Science and Technology*. ed. by R.A. Meyers (Springer New York, New York, NY, 2018), pp.1–16
4. S. Yadav, A. Sharma, Importance and challenges of hydrothermal technique for synthesis of transition metal oxides and composites as supercapacitor electrode materials. J. Energy Storage **44**, 103295 (2021)
5. T.N. Phan, M.K. Gong, R. Thangavel, Y.S. Lee, C.H. Ko, Enhanced electrochemical performance for EDLC using ordered mesoporous carbons (CMK-3 and CMK-8): role of mesopores and mesopore structures. J. Alloy. Compd. **780**, 90–97 (2019)
6. T.S. Le, T.K. Truong, V.N. Huynh, J. Bae, D. Suh, Synergetic design of enlarged surface area and pseudo-capacitance for fiber-shaped supercapacitor yarn. Nano Energy **67**, 104198 (2020)

Chapter 2
State-of-the-Art Works in Developing Supercapacitor

2.1 Current Development of Transition-Metal-Based Supercapacitors

Considerable research has been devoted to critical Ni-based materials to enhance the performance of supercapacitors, as seen in Fig. 2.1 [1–4]. The current progress is introduced with a discussion of significant outcomes that provide support for the objectives of the proposed effort. Su et al. designed a $Ni(OH)_2$@Ni core–shell electrode with a Cs of 2454 F/g at 5 A/g current density and excellent cycling stability 11 for high-performance and flexible SC. Miao et al. introduced a stable NiS@CoS core–shell electrode into their study. During the testing process, the electrode exhibited a capacitance retention of 82% at 10 A/g and a Cs of 1210 F/g at 1 A/g [5]. Wan et al. grew nanosheet arrays of $NiFeP$@$NiCo_2S_4$ on carbon cloth for asymmetric SC with the obtained Cs of 874.4 F/g at 1 A/g and capacity retention at 85.6% after 5000 cycles [6]. Lu et al. fabricated a hybrid film of Ni–Co sulfide nanosheet/CNT for asymmetric flexible SC with an energy or power density of 900 or 18000 W/Kg and a voltage output of 1.8 V. Capacity retention of 80.64% is achievable after 10,000 cycles [7]. NiS significantly increased the specific capacity of MoS_2 nanosheet arrays while maintaining their exceptional flexibility and charging-discharging stability, as discovered by Guan et al. [8]. The results of those works revealed that Ni-based materials are good choices for cathode electrodes with relatively high electrochemical performance.

The other famous V-based SCs have also been widely studied for batteries and SCs due to their unique layered structure with various valences, low cost, and abundance [9]. However, the performance of V-based energy materials is constrained by factors such as an unstable charge–discharge process, suboptimal electrochemical, and cycling capabilities [10]. Some approaches have been adopted to compensate for the limitation, as mentioned above. Wei et al. incorporated V_2O_5 with N-doped graphene aerogel and a specific capacity of 710 F/g (at 0.5 A/g) was obtained with

© The Author(s), under exclusive license to Springer Nature Singapore Pte Ltd. 2024
H. Abdullah, *Vanadium Oxide-Based Cathode for Supercapacitor Applications*,
SpringerBriefs in Applied Sciences and Technology,
https://doi.org/10.1007/978-981-97-5243-0_2

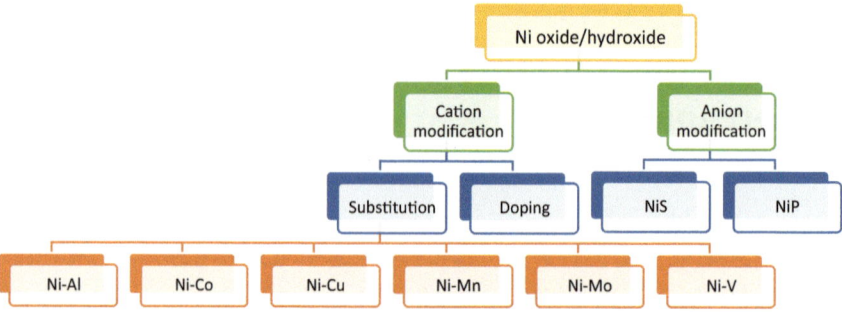

Fig. 2.1 Nickel-based cathode materials for SCs employing various methodologies [1–5]

good cycling behavior (95% capacity retention after 20,000 cycles) [11]. Balamu-rugan et al. fabricated vanadium nitride/N-doped graphene with stable capacity reten-tion at 445 F/g, which reached 99% after 10,000 cycles at 10 A/g current density [12]. Wang et al. formed a composite of V_2O_5 nanoflakes and Ni_3S_2; after 2500 cycles, the composite achieved a maximum Cs of 3060 F/g with an 85% capacity retention [13]. Another strategy from Shakir et al. utilized a thin layer of V_2O_5 on MWCNT with graphene composites to induce high capacitance at 2590 F/g and rela-tively stable capacity retention of 96% after 5000 cycles [14]. As vanadium-based materials are also used to advance battery performance, some properties of those materials are advantageous for battery-type SC applications [15]. In addition, since MO-based SCs exhibit comparatively superior performance, as shown in Fig. 2.2, the capacitance can be further increased by combining the V-based SC with other metal-oxide (MO) materials.

Moreover, to increase the energy storage capacity of LIBs, a substantial amount of research has been devoted to the investigation of anode materials composed of transi-tion metal oxides (TMOs) [17, 18]. Transition metal oxides are practical anode substi-tutes due to their exceptionally high specific capacities during the conversion mech-anism of a chemical reaction [19]. The exceptionally high capacity of these materials (700–1200 mAh/g) is equivalent to three times that of commercial graphitic carbon. Unfortunately, the electrical conductivity of these systems is comparatively low, and they inevitably experience a significant hysteresis between the lithiation and delithi-ation processes; this renders them unusable [20]. Poor conductivity always results in inferior rate performance and power density. As the volume expands, pulverization occurs, resulting in poor lithium storage performance and rapid capacity degrada-tion. TMOs have undergone extensive efforts to enhance their surface area, mechan-ical strength, mass and electron transport kinetics, electrical conductivity, and elec-trochemical kinetic properties in order to overcome these deficiencies. The design of nanosized structures of different polymorphs is adopted, such as nanoparticles, nanorods, nanosheets, nanowires, and nanotubes [21, 22]. Nanomaterials with small size effects can shorten the diffusion path for Li^+, while surfaces with large specific surface areas can provide more sites for Li^+ insertion. TMOs and high-conductive materials can substantially improve performance rate and electronic conductivity, in

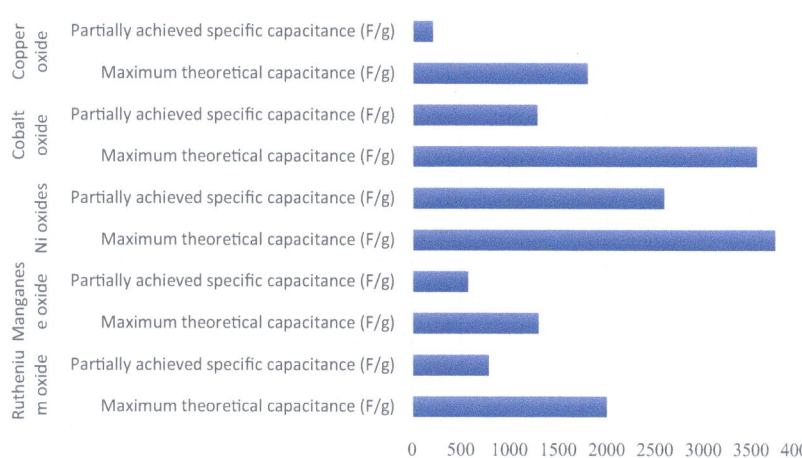

Fig. 2.2 Specific capacitance of metal oxide-based SCs synthesized via electrodeposition [16]

addition to facilitating the design of suitable structures. Carbon, carbon nanotubes (CNTs), graphene, reduced graphene oxide (RGO), multiwalled carbon nanotubes (MWCNTs), etc. are among the highly conductive materials and also possible for the application to overcome the TMO limitation [23–26].

Developing hybrid devices to combine battery and supercapacitor is an exciting idea to improve the energy storage device performance diagonally toward gasoline-based fuels in the Ragone plot. Based on the literature reviews, the vanadium-, nickel-, mangan-, and cobalt-based electrode materials are promising to be developed for high charge storage capacity. Transportation facilities such as cars, trains, buses, and other machines require high-performance energy storage technology with a fast and highly reversible charge–discharge process. Therefore, the exploration using vanadium-based cathode as the battery materials will be an advantage in providing electrodes with high energy density. Modification with TMO- and nickel-based materials will enhance the power density due to the basic property of pseudocapacitor.

2.2 Design Strategies in VO$_x$-Based Electrodes

The accumulation of charge on the electrode surface or electron gain and loss followed by ion insert/extract are the primary mechanisms by which charge is stored in an electrical energy storage system [27]. These processes all favor electrode materials characterized by high specific surface area and conductivity. In general, the characteristic attributes of transition metal oxides (TMOs) that enable the intercalation of ions and electrons into the lattice of metallic compounds [28–31], should be identified as the fundamental cause. To attain the desired performance characteristics of energy storage devices, it is critical to possess the ability to consistently generate and optimize V-compounds in accordance with particular requirements. Successful design strategies for property modulation are summarized in Fig. 2.1. The strategies to improve VO_x-based SCs are tuning the morphology, structural and electronic, and optimizing their synergistic effects.

2.2.1   Morphological Tuning

An anomalous behavior exhibited by nanostructures is explained by Fick's law ($t \sim X^2/D$), which states that a reduction in diffusion length (X) results in a decrease in ion diffusion time (t). This particular occurrence provides evidence in favor of the notion that nanostructuring facilitates an increased influx of ions across the interface, thereby accelerating the reaction rate of vanadium-based electrodes. A pseudocapacitive characteristic becomes more evident when the dimensions of V-compounds approach the nanoscale, which is beneficial in mitigating the slow solid-state diffusion of intercalating ions [32]. By accumulating a substantial number of charges in the vicinity of the surface in the microsecond to millisecond time range, the reaction taking place in nanostructured electrodes effectively inhibits intercalation-induced phase transition, resulting in superior kinetics compared to the conventional diffusion-controlled faradaic reaction. This hypothesis states that the creation of a capacitive controlled storage mechanism can account for the drawback-indistinguishable rate capability in nanostructured V-compounds [33]. Some important strategies are shown in Fig. 2.3 to improve the electrochemical performance of VO_x-based electrodes for SCs.

2.2.2   Structural and Electronic Tuning

As demonstrated by kinetics research, the lattice environment has a substantial impact on the way in which cations interact with host materials. Doping is an exceptionally effective strategy for effecting change. It injects metal ions or nonmetallic heteroatoms into the structure of electrode materials via in-situ or ex-situ processes. To improve electrochemical performance, the electronic distribution of electrode materials can be modified via doping with ions or atoms [34]. In order to rectify the state of electrical neutrality imbalance within the crystal lattice, doping cations is utilized to introduce imperfections, including vacancies and holes [35]. Hole formation can consequently lead to an enhancement in electronic conductivity. The classic Arrhenius equation illustrates the correlation between the diffusivity of ions and the energy required for activation. $D = D_0.\exp(-\Delta G/K_B.T)$; ($\Delta G$ is the energy barrier, K_B is the Boltzmann constant, T is the temperature, and D_0 is the preexponential

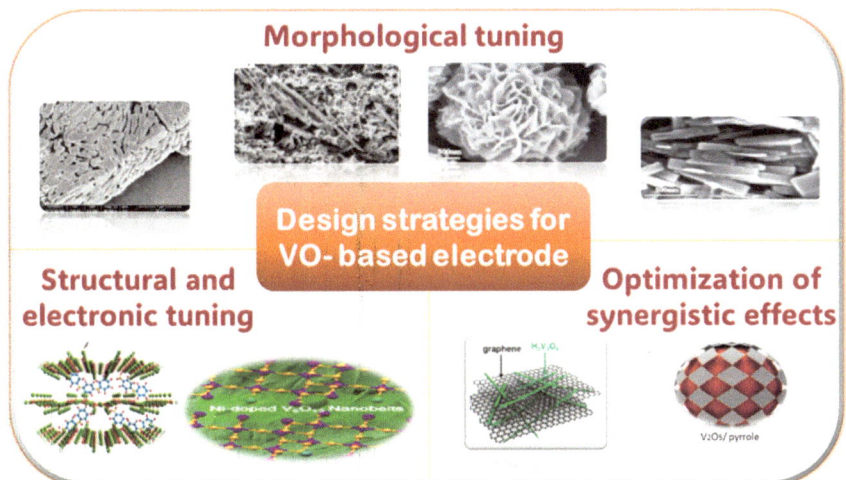

Fig. 2.3 Design strategies in the electrochemical performance of VO$_x$-based electrodes for SCs [46]. (Elsevier Copyright 2024 with license number: 5772230848288)

factor) [33]. To adjust the activation energy of *V*-compounds, lattice engineering through ion doping and lattice expansion might possibly be the preferred approach.

Conversely, in order to expand the interlayer space, preserve the structure, and increase the reversible capacity, cations and micromolecules may function as pillars. With the intercalation of cations between the layers, the electrochemical environment between the V–O layers can be modified, thereby affecting the distribution of electron clouds in the V–O layer. Consequently, doping foreign atoms into vanadium oxides can enhance their electrical and ionic conductivity [36, 37]. At present, the appropriate cations for intercalation are ions of alkali metals, transition metals, and alkaline-earth metals. Variations in valence state, ion radius, and number of inserted ions result in structural differences among the intercalated products. Different distortions are introduced into the single layer by cations with varying ion radii during the intercalating process [37, 38]. The difference in binding energies between oxygen atoms and cations in the V–O layer further contributes to structural modifications. Notwithstanding this, it is crucial to underscore the fact that metal ions characterized by substantial interlayer space occupation, wide ionic radii, and high molar masses (e.g., Cu and Ag) can result in suboptimal specific capacities and energy densities [27].

An additional effective doping method involves the utilization of non-metallic heteroatoms. Adding substances such as fluorine, nitrogen, phosphorus, sulfur, or sulfur dioxide to conductive polymers or carbon-based materials compounded with vanadium oxide-based materials is a common method of non-metallic heteroatom doping. The bonding force between the composite materials is significantly increased through the doping of elements with strong electronegative valence, resulting in a substantial improvement in the cycle stability and rate performance of the electrode

[39]. In recent times, conductive polymers such as polyaniline (PANI), polypyrrole (PPy), and poly(3,4-ethylenedioxythiophene) (PEDOT) have been introduced into the interlayer of vanadium oxide-based materials as guest species [40]. The addition of a conductive polymer improves cycle performances and rates by enhancing the electrode's capacity to transport ions and electrons [41, 42].

2.2.3 Optimizing Synergistic Effects

Through the regulation of chemical reactivity and surface chemistry, surface engineering improves conductivity and structural integrity. Conformal carbon coating is widely used for V-compounds due to its ability to offer several advantages in a straightforward manner through the utilization of carbon sources that are easily obtainable, including polymers, polysaccharides, and organic acids [43]. The enhanced electrochemical capabilities are due to two factors: (i) an increase in electronic and ionic conductivity, and (ii) mechanical confinement, which mitigates excessive volume expansion. Addressing the sluggish kinetics of electrical and ionic transport observed in the majority of V-compounds can be accomplished effectively by implementing a synergistic coupling effect via the juxtaposition of different material types. In order to maximize the potential of V-compounds, carbon-based scaffolds can provide them with exceptionally suitable properties, among other possible combinations [3]. For example, the enhancement of electrochemical properties in carbonaceous materials resulting from the incorporation of V_2O_5 [44], $Li_3V_2(PO_4)_3$ [35], and $Na_3V_2(PO_4)_3$ [45]. This improvement can be attributed to the subsequent factors: Carbon supports with high conductivity facilitate efficient charge transfer, thereby averting significant agglomeration and safeguarding the structural integrity of V-compounds via carbon confinement. Furthermore, carbon exhibits a remarkable level of tunability, enabling it to transform into an extensive array upon the introduction of defects and vacancies caused by doped or absent atoms. The investigation into conductive scaffolds comprised of organic materials with intrinsic doping (e.g., polymers and ionic liquids), chemically produced doped frameworks, and materials derived from living organisms has been spurred by these discoveries. The carbon that has been modified has enhanced wettability on its surface and cation-friendly binding sites. The enhanced reaction kinetics may be attributed to these unique characteristics because of the synergistic enhancement of the electrode's electronic conductivity facilitated by carbon materials.

2.3   Recent Progress on VO_x-Based Electroactive Materials

Material selection, identification of synthesis methods, characterization, fabrication, and electrochemical studies are examples of general experimental steps involved in the development of V_2O_5-based electroactive materials suitable for supercapacitor applications (Fig. 2.4) [47]. The synthesis procedures significantly influence the material morphology and properties. The capacitance of nanoporous V_2O_5 produced via sol–gel synthesis was 214 F/g, whereas co-precipitation at the same electrolyte concentration yielded 349 F/g. This phenomenon unequivocally demonstrates how the synthesis method affects the electrochemical properties of material [47]. To date, researchers have investigated various morphologies and structures of electrode materials based on V_2O_5, including nanorods [48], nanoflowers [49], nanobelts [50], nanowires [51], nanofibers [52], hierarchical spheres [53], nanosheets [54], and nanotubes [55]. These investigations aimed to demonstrate how morphologies affected the electrochemical properties and supercapacitive performance of V_2O_5-based electroactive materials, ultimately with the objective of improving the materials' electrochemical functionality. Variation in the morphology of the material can modify the properties of the material. This section presents a review of the synthesis and supercapacitive performance of V_2O_5 electroactive material having different morphologies. For instance, Patil et al. [56] synthesized V_2O_5 powder through a thermal decomposition process and deposited it on a Ni foam substrate. The material reserved 96.1% of its specific capacitance (1227.2 F/g) at 1 mV/s after 2000 GCD cycles. Li et al. [54] synthesized V_2O_5 nanosheets through the hydrothermal method and exhibited 96.8% cycling capacity retention with a specific capacity (375 F/g) than V_2O_5 particles (318 F/g) in K_2SO_4 solution. The nanosheet retained 92.6% of its specific capacity over 500 cycles. Yang et al. [57] prepared hollow V_2O_5 spheres that provided 479 F/g specific capacitance. The core finding of such studies has confirmed the impact of synthesis method on the morphology and electrochemical properties of the materials. However, none of the studies provide a consistent pattern for how the morphology of material influences its electrochemical performance.

Furthermore, Zheng et al. [58] investigated the morphology of V_2O_5 microcrystals in the form of rhombohedral, butterfly-like, and flower-like structures. The microcrystals demonstrated cycling stabilities that fluctuated nearly consistently over 2000 cycles, with specific capacitances of 641, 556, and 609 F/g at 0.5 A/g; their alterations were 119.8%, 132.6%, and 70.4%, respectively. The synthesis methodology employed to attain the desired morphologies of the products is depicted in Fig. 2.4a with XRD analysis in Fig. 2.4b. Each product demonstrates a high degree of crystallinity and crystallization, as illustrated by the XRD plot. It entails the adjustment of critical parameters, such as the temperature, volume of ethanol, and concentration of NH_4VO_3, throughout the crystallization procedure. Comparable Raman spectra were observed for each of these morphologies, as documented by the authors (Fig. 2.4c). The morphological distinction (see Fig. 2.4d) is essential for regulating the dissolution or volume expansion of V_2O_5 during charging and discharging in aqueous electrolytes, which leads to poor cycling stability. As a result of electrode polarization, the

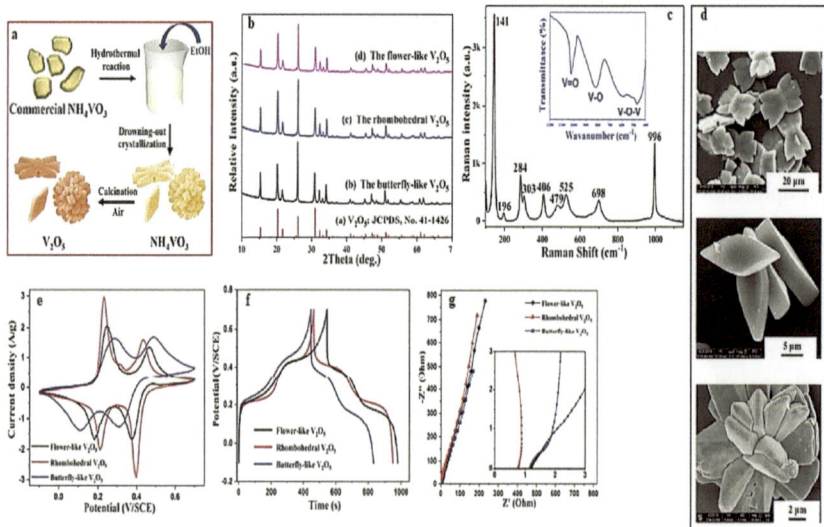

Fig. 2.4 a V$_2$O$_5$ synthesis process with **b** XRD patterns and **c** Raman spectra of **d** SEM images of butterfly, rhombohedral, and flower-like V$_2$O$_5$ and additional comparison of electrochemical properties across various morphologies of V$_2$O$_5$: **e** CV plot at 1 mV/s; **f** GCD plot at 1 A/g, and **g** Nyquist plots within range of 100 to 0.01 Hz. (Reprinted with permission from ref. [58]. Copyright 2018 American chemical society)

oxidation and reduction peaks were displaced to higher and lower potentials, respectively. Identifying redox peaks in the CV curves, as depicted in Fig. 2.4e, confirms that the electrode material possesses pseudocapacitor properties, as supported by the data in Fig. 2.4f and g. As a result, using morphologically controlled materials is commonly advised and mandatory in developing high-performance supercapacitors [58].

2.4   Regulation and Control Methods to Improve Electrochemical Properties of Electrodes

Due to their numerous active sites, high aspect ratio, large specific surface area, and ease of formation into three-dimensional networks, nanowire materials have garnered considerable interest in energy storage and other sectors, among the numerous nanomaterials. In contrast, the electrode material structure is detrimentally affected throughout the performance cycle due to the persistent insertion and extraction of ions. This results in self-aggregation and reduced conductivity. To address the issues of structural deterioration and poor stability throughout the cycle, it is critical to regulate the performance of nanowire electrode materials. The aspects of electronic structure regulation, surface/interface regulation, ion diffusion channel regulation,

ion/electron dual continuous regulation, and structural stability control are primarily addressed in this section related to the regulation and control methods of the electrical transport performance of high aspect-ratio electrode materials.

2.4.1 Electronic Structure Optimization

Analysis of the electronic state and its motion characteristics in crystals frequently employs the energy band theory, including metal, insulator, and semiconductor. Electrode materials consisting of nanowires are frequently subjected to additional fields, ionic pre-intercalation, and elemental doping to enhance their electric conductivity and transport capabilities.

2.4.1.1 Ionic Pre-intercalation

Ionic pre-intercalation is a common technique used to enhance the electric transport performance of layered crystals, such as α-MoO_3, σ-MnO_2, and V_2O_5 by pre-intercalating ions with electrochemical transport capabilities, including Li ions, Na ions, and Ca ions, into interlayers. This process optimizes the band structure of the crystals. δ-$Na_{0.33}V_2O_5$ ZIBs anode materials were synthesized by Mai group through chemical pre-intercalation of sodium ions into V_2O_5 [59]. The sodium ions were found to be intercalated between layers of $[V_4O_{12}]$ as seen in Fig. 2.5a. Conductivity measurements were conducted on individual nanowires both prior to and subsequent to the pre-intercalation of sodium ions. The results indicate that the electronic conductivity of the sodium pre-intercalated δ-$Na_{0.33}V_2O_5$ nanowires is four orders of magnitude higher than that of the α-V_2O_5 nanowires, increasing from 7.3 Sm^{-1} to $5.9 \cdot 10^4$ Sm^{-1}. This substantial enhancement significantly optimizes the electric transport performance. In the context of ZIBs, these materials demonstrate exceptional rate performance and stable cycling behavior, as evidenced by a capacity retention of more than 93% after 1000 cycles. Utilizing an iron pre-intercalation approach, Mai group additionally documented a novel and uncomplicated method to impede the lattice breathing of vanadium oxide xerogel during sodiation/desodiation [60]. Lattice breathing along the c-axis of VO_x is significantly diminished from 3.74 to 0.49 Ω by Fe pre-intercalation, as seen in Fig. 2.5b. Additionally, for reversible Na^+ insertion/extraction, pre-intercalation of iron results in increased stabilized interlayer spacing. The Fe-VO_x demonstrates improved cycling and rate performance due to the suppression of lattice breathing and the stabilization of interlayer spacing.

2.4.1.2 Elemental Doping

Elemental doping is a process in which a small amount of another element is intentionally doped into a material to enhance its performance. Doping can impart specific

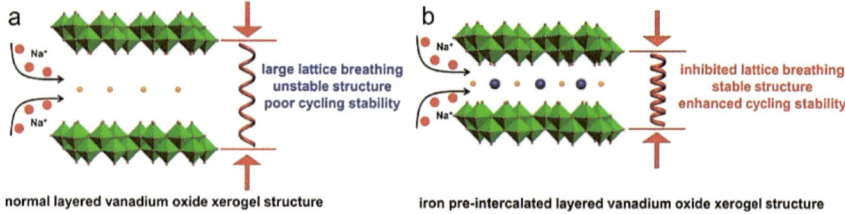

Fig. 2.5 Schematic illustration of layered VO_x structure for **a** normal and **b** iron pre-intercalation structures. (Reprinted with permission from ref. [60]. Copyright 2015 American chemical society)

electrical, magnetic, and optical properties to materials, endowing them with particular merits or applications. According to reports, introducing impurity levels into the lattices of materials through doping can effectively alter the electronic structure of the lattice. This alteration increases the concentration of charge carriers, which enhances the electric transport performance of the material without compromising its structural stability [61, 62]. So far, this principle has been implemented in the system of multivalent ion batteries. By doping TiO_2, Damboumet et al. generated a substantial quantity of titanium vacancies, thereby increasing the storage capacity of Mg^{2+} and Al^{3+} [63]. Additionally, doping Ag^+ into α-MnO_2 has been documented to generate vacancies, providing further evidence that doping metallic elements into a material can be advantageous for adjusting its electrochemical properties and introducing vacancies [64]. Mai's group conducted an investigation into the surface gradient titanium doping of MnO_2 nanowire (Ti-MnO_2) as a prospective cathode material for aqueous zinc-ion batteries [65]. The doped Ti-MnO_2 exhibits a reduced charge transfer resistance compared to unmodified MnO_2. DFT calculations indicate that the oxygen vacancy and Ti substitution result in the formation of a charge depletion zone and an inherent electric field. The interfacial electric field and oxygen vacancies generated by the unbalanced charge and spin distribution in the tunnel surrounding the Ti substitution site may facilitate the opening of the [MnO_6] octahedral walls, thereby accelerating ion diffusion and electron transport kinetics.

2.4.1.3 Introducing Extra Field

The electrostatic field between charged particles is a universal force. Particularly in electrochemical devices based on pseudocapacity, an electrical double layer on the surface of the electrode material significantly contributes to the energy storage reaction. We can optimize the kinetics and thermodynamics of an electrochemical reaction if it is feasible to manipulate and regulate ions at the electrochemical interface and charge transport within electrode materials by applying an additional electrostatic field to the interface of the electrochemical reaction. By applying a back-gate voltage to the pristine VSe_2 nanosheet, Mai group reported a method to tune the dynamics of the adsorption process in HER [66]. Redistribution of ions at the interface between the electrolyte and VSe_2 nanosheets is induced by the back-gate voltage; this facilitates the rate-limiting step (discharge process) under HER conditions and enables enhanced electron transport. VSe_2 nanosheets manage to attain a significant low-onset overpotential of 70 mV in the absence of any chemical treatment. The VSe_2 nanosheet's field-tuned adsorption dynamics are responsible for this unexpected improvement, as evidenced by the significantly reduced charge transfer resistance (from 1.03 to 0.15 MΩ) and adsorption process time constant (from 2.5×10^{-3} to 5.0×10^{-4} s).

2.4.2 Optimization of Interfaces

The electrochemical performance of vanadium-based electrode materials is frequently influenced by the surface/interface structure, affecting charge transfer and ion storage. Consequently, vanadium-based electrode materials feature distinct surface/interface structures that manifest such variations. Modifying the surface and interface structure of electrode materials based on vanadium mainly involves compounding conductive materials and structural design.

2.4.2.1 Utilizing Conductive Materials in Composition

To some degree, the issue of unstable interface structure of nanowire and buffer volume expansion can be effectively resolved by combining conductive material with electrode material. Concurrently, the advancement of electrical conductivity can be substantially enhanced through the incorporation of conductive material, thereby facilitating the swift transport of electrons. Conductive carbon materials (graphene [67], carbon nanotubes [68], amorphous carbon [69]), conductive polymers (polypyrrole [70], polyaniline [71]), and conductive hybrid materials (polyaniline [71]) are frequently employed conductive materials. To optimize the electrochemical properties of these materials, vanadium-based electrode materials are utilized in the synthesis process. The study conducted by Mai group involved the construction and investigation of the lithium storage performance of a mesoporous carbon-coated $Li_3V_2(PO_4)_3$ nanowire structure [72]. Mesoporous carbon uniformly coats $Li_3V_2(PO_4)_3$ nanocrystals, and the nanowire structure of $Li_3V_2(PO_4)_3$ effectively shortens the migration path of Li^+. The carbon-coated mesoporous structure improves electron transport in the electrode material, mitigates the local volume expansion induced by Li^+ insertion during cycling, and contributes to the structural stability of $Li_3V_2(PO_4)_3$. Extra ion storage sites may also be provided by interfacial nanocavities produced via van der Waals or heterobond interactions. Cathode material composed of interfacial V–O-C bonds (Fig. 2.6) was fabricated by Mai group using VO_x sub-nanometer cluster/reduced graphene oxide (rGO) [73]. In contrast to conventional mechanisms, Zn^{2+} ions are demonstrated to be primarily stored at the interface between VO_x and rGO. This phenomenon induces typical valence state transitions and takes advantage of the storage capability of the exceptionally conductive

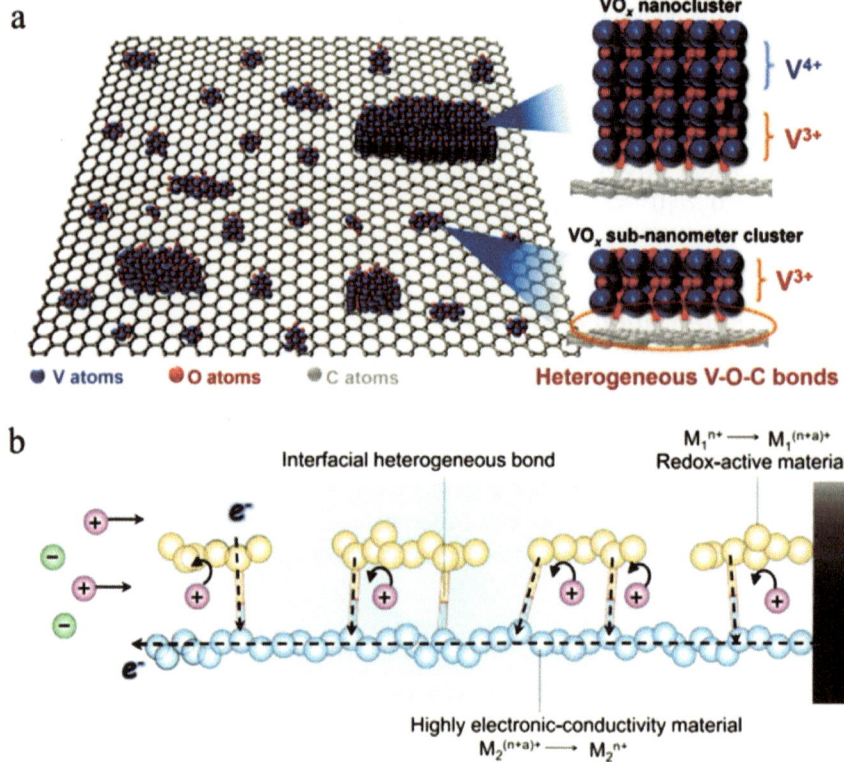

Fig. 2.6 a Heterogeneous V–O–C bonds and **b** the interface pseudocapacitance [74]. (John Wiley and Sons 2021 with license number: 5772251186417)

rGO without requiring energy storage activity. Additionally, the reversible destruction/reconstruction of this interface enables it to store a greater quantity of ions compared to the bulk phase, and it induces decoupled electron/Zn^{2+} transport. High specific capacity has been achieved by the cathode material, as demonstrated by the results (443 mAh g^{-1} at 100 mA g^{-1}, which exceeds the theoretical capacity of various surface components). This storage interface is an innovative method for constructing devices with high energy and power densities.

2.4.2.2 Constructing Nanostructures

By facilitating ion migration and electron transport, as well as promoting electrolyte infiltration, three-dimensional, porous, or ultra-thin electrode materials that possess a high specific surface area can enhance the electrochemical performance of vanadium-based electrode materials. A universal liquid phase stripping technique was utilized by Mai group to fabricate a three-dimensional $H_2V_3O_8$ hydrogel structure comprised of self-curling nanowires and ultrathin nanoribbons [75]. By cross-linking ultrathin nanoribbons and self-crimping nanowires, a hydrogel structure is produced that drastically reduces the ion diffusion distance. By increasing the contact area between the electrode and electrolyte, the porous hydrogel structure creates an efficient interconnecting channel that promotes ion diffusion. Furthermore, the construction of ultrathin pre-lithiated V_6O_{13} nanosheet cathodes [76] and hierarchical three-dimensional porous V_2O_5 microplates [77] is also undertaken.

The practical implementation of solid-state lithium batteries (SSLBs) is impeded by interfacial chemical reactions and inadequate solid–solid interfacial contact, specifically near the cathode. Mai group devised a gradient $H_2V_3O_8$ nanowire (NW) cathode [78] in order to tackle this issue. Two interfacial buffer layers composed of polyvinyl oxide (PVO) are utilized to form gradient distribution interfaces between the two surfaces of this distinctive gradient electrode. The more ionic-conductive polymer establishes a seamless interface with a solid–solid electrode (SSE) on one side of its surface, whereas the more electron-conductive $H_2V_3O_8$ NWs/rGO facilitates rapid electron transport in its capacity as a collector fluid on the other. The internal NW cathode material is also coated uniformly with a solid polymer electrolyte composed of rGO and PVO. By transforming point-to-point contact between the positive electrode and the SSE into large-area contact, this strong bonding improves structural stability and provides a continuous channel for rapid electron/ion transport [79–81].

2.4.3 Ion Diffusion Channel

A critical parameter that reflects the ion transport properties of substances is the ion diffusion coefficient. Determining the distance between lithium layers and electrostatic repulsion is a critical influencing factor [82]. Ceder group systematically investigated the migration characteristics of lithium ions in layered transition metal oxides using a DFT theoretical calculation. The interdependence between influential factors is an electrostatic interaction force that also impacts the layer spacing. Electrostatic interaction force is calculated as follows, in accordance with Coulomb's law: $u = \frac{K_e q_1 q_2}{r^2}$, in which q denotes charge, while r signifies the separation between the two charges. The formula uses k_e to denote the coulomb coefficient ($k_e \approx 9.0 \times 10^9$ Nm^2/C^2). The resistance of current-carrying ions in the pre-intercalated material is diminished as the distance r increases due to the expansion of layer spacing along the c-axis. It results in a decrease in the electrostatic force (u). The ion diffusion coefficient of materials and their electrical transport properties can be optimized and enhanced by modifying the diffusion channel of electrochemical transport ions. To increase the ion transport capacity of nanowire electrode materials, this section will describe how ion diffusion channels can be regulated via pre-intercalation of organic molecules, pre-embedding of metal ions, and pre-intercalation of inorganic/nonmetallic ions [60, 83–88].

2.4.3.1 Metal Ion Intercalation

To improve the electrochemical properties of the materials, Mai group investigated the pre-intercalation of alkali metal ions into the typical nanowire cathode material A–M–O (A = Li, Na, K, Rb; M = V, Mo, Co, Mn, Fe) and quantitatively described the process [85]. The pre-intercalation of sizable alkali metal ions can enlarge the diffusion channel of lithium ions in lithium-ion batteries. The results indicate that vanadium oxide nanowire electrode material with pre-intercalated alkali metal ions exhibits remarkable rate performance. The dimension of the ion diffusion channel after the pre-intercalation of an assortment of distinct ions was calculated with experimental characterization (X-ray diffraction and rotational electron diffraction) and DFT theory. Further analysis yielded the subsequent findings: Particularly in the cathode material of lithium-ion batteries, potassium ion pre-intercalated vanadium oxide exhibits the greatest interlayer distance and the highest multiplier performance; thus, expanding the ion diffusion channel aids in enhancing the Li-ion diffusion performance. A pre-intercalated vanadium oxide (δ-$K_xV_2O_5 \cdot nH_2O$) with a pre-intercalated layer spacing of 9.56 Å was developed by Pomerantseva et al. [89] from Drexel University. The specific capacity of δ-$K_xV_2O_5 \cdot nH_2O$ is notably high at 268 mAh g^{-1} when employed as the cathode material in a potassium-ion battery. This result is attributed to the substantial layer spacing of δ-$K_xV_2O_5 \cdot nH_2O$, which facilitates the transport of potassium ions.

The investigation of alkaline-earth metal ion pre-intercalation has garnered significant attention, alongside the study of alkali metal ion pre-intercalation. The authors of the study conducted a comparison of vanadium oxides (δ-$M_xV_2O_5$, M = Li, Na, K, Ma, Ca) pre-intercalated with various alkali metal ions and alkaline-earth metal ions. Their investigation established the relationship between the layer spacing and the hydration radius of the pre-intercalated ions [87]. Through the synergistic interaction of Mg^{2+} and lattice water, Mai group successfully synthesized the double-layer structure $Mg_{0.3}V_2O_5 \cdot 1.1H_2O$. Notably, the electrode material achieved exceptional cycle stability and high conductivity by pre-intercalating Mg^{2+} ions [90]. The charge shielding effect of the lattice water can accelerate the migration rate of Mg^{2+}. It exhibits exceptional cycle stability and rate performance when utilized as the cathode material in magnesium-ion batteries (the specific capacity retention rate is 80% after 10,000 cycles). Further, pre-intercalation materials may contain additional metal ions, including Mn^{2+}, Fe^{3+}, and Al^{3+}. By pre-intercalating iron ions, Mai group could control the ion channel of vanadium oxide [60]. Fe–VO_x vanadium oxide nanoribbons were synthesized using a straightforward hydrothermal method. Crystal structure instability can result from the pre-intercalation of iron ions, which inhibits the "lattice breathing" produced during sodium ion de-embedding. The cyclic stability of sodium ion storage can be enhanced by mitigating the collapse of the diffusion channel caused by this method. The charge transfer impedance is diminished, whereby the diameter of the half circle signifies a reduction in the charge transfer performance of vanadium oxide pre-intercalated with sodium ions. Specific capacity and cycle stability metrics are enhanced when sodium-ion batteries employ the material as the anode.

2.4.3.2   Organic/Inorganic Ion Intercalation

Pre-intercalation of organic molecules/nonmetal ions (e.g., NH^{4+}) is a viable approach to enhance and optimize the ion diffusion channels of electrode materials. Yang et al. and Mai group collaborated to develop a vanadium pentoxide gel ($V_2O_5 \cdot nH_2O$) pre-intercalated with water molecules [88]. This gel was implemented as the cathode material in a zinc-ion battery. Throughout the discharge procedure, zinc ions became encapsulated within the $V_2O_5 \cdot nH_2O$ layers. The electrostatic interaction between zinc ions and V_2O_5 is diminished, the layer spacing is increased, and the effective charge of zinc ions is decreased through solvation with pre-intercalated water molecules. The functioning of this process can be described as follows: pre-intercalated water molecules serve as a "lubricant" in the zinc ion diffusion process, thereby facilitating efficient zinc ion diffusion and culminating in an exceptionally effective electrochemical zinc storage performance. To increase layer spacing, Kanatzidis et al. use the atomic pair distribution function (APDF) to separate and stack water molecules in an infinite stack along the Z axis [83]. Consequently, the performance of this V_2O_5 gel was superior to that of pristine vanadium pentoxide.

2.4.3.3   Organic Molecule Intercalation

Pre-intercalation of organic and inorganic molecules (e.g., pyridine, triglycol, aniline) and metal ions into the crystal structure of nanowire electrode material can further optimize the ion diffusion channel. In their study, Yu et al. implemented a pre-intercalation technique using the organic molecule triethylene glycol to augment the layer spacing of the layered polyanionic $VOPO_4$ electrode material. The resultant sodium-ion battery electrode material was $VOPO_4$ pre-intercalated with triethylene glycol. In contrast to $VOPO_4$ lacking pre-intercalated triethylene glycol, the presence of sodium ions resulted in a substantial reduction in the diffusion energy barrier, thereby substantially enhancing the sodium ion diffusion kinetics [86]. By pre-intercalating aniline molecules into the $VOPO_4$ crystal, Mai group increased the layer spacing of $VOPO_4$, which is advantageous for polyvalent metal-ion batteries and facilitates the transport of ions with a large radius [91].

Furthermore, Mai group investigated the sodium storage capabilities of $V_2O_5 \cdot nH_2O$ nanowires that had been pre-intercalated with pyridine molecules [92]. Pyridine molecules exhibit a high propensity for interlayering, proton fusion, and hydrogen bond formation between layers of bilayer $V_2O_5 \cdot nH_2O$ owing to the considerable interlayer spacing and unimpeded proton mobility between layers. Through the implementation of this methodology, it is possible to consistently uphold the interlayer structure in a stable condition and avert its collapse, thus ensuring the protection of the interlayer transport channel for sodium ions. The exceptional stability and high specific capacity of $V_2O_5 \cdot nH_2O$ pre-intercalated with pyridine in sodium-ion batteries have been demonstrated.

2.4.4 Ion/Electron Dual Continuous Regulation

Electrical transport of substances is constrained by the velocity at which ions and electrons are moved. When it comes to electrode materials, however, the process of optimizing electrical transport usually begins unilaterally with ions or electrons, despite the negative impact of the unoptimized component. By strategically designing the coaxial semi-hollow structure and the nanowire array structure, it is feasible to control the electrical transport performance in the nanowire electrode material to attain bi-continuous electron/ion transmission.

2.4.4.1 Semi-Hollow Structures

To optimize electronic conduction, it is possible to apply a dense coating of conductive carbon, graphene, conducive polymer, or other conductive substances onto the active material. However, the tight coating will hinder the ability of the electrode material to facilitate the efficient transfer of ions and electrons by obstructing the movement of ions in the electrolyte through the material. The coaxial semi-hollow structure, as opposed to tightly coated nanowires, enables thorough infiltration of the electrolyte via the semi-hollow structure and simultaneously establishes a swift electron conduction pathway through the outer conductive material. In accordance with the formula description [79], Mai group successfully synthesized a V_3O_7/graphene material characterized by a coaxial semi-hollow nanowire structure. This configuration enabled the bi-continuous transport of electrons and ions. By integrating experimental data and molecular dynamics simulations, they proposed a growth mechanism that combines "self-curling" and "oriented assembly." This involves the introduction of a precise amount of energy into the synthesis environment to cause graphene on the surface of V_3O_7 nanowires to curl inwards. Through the establishment of a distinct interstitial region between V_3O_7 nanowires and graphene, this approach produces coaxial semi-hollow nanowires that enable the bi-continuous movement of ions and electrons. Furthermore, apart from enhancing electrical conduction, this arrangement produces an empty vacuum region positioned between the nanowires and the graphene rolls. By rendering the interior of the semi-hollow structure entirely permeable to the electrolyte, the interface contact between the electrode material and the electrolyte is strengthened. As a consequence, polarization is diminished and charge distribution becomes more uniform throughout the energy storage process. The interface charge transfer resistance of the coil coaxial semi-hollow V_3O_7/graphene nanowires is reduced from 81 Ω to 32 Ω, as determined by the charge transport performance test, when compared to the conventional V_3O_7 nanowires. The conductivity of these nanowires also increases by a factor of 27. The constructed energy storage device not only sustains a substantial energy density but also increases its power density by a factor of six. Furthermore, the cavity possesses the capacity to accommodate the volume expansion of the material, guaranteeing the continuous transport of electrons and ions throughout the entire cycle.

Furthermore, to verify the previously described ideas, Mai group fabricated MnO_2 nanorods that exhibited a coaxial semi-hollow structure [80]. Utilizing the electron/ion bi-continuous effect to modulate the kinetics of the conversion reaction in a negative electrode material system was discovered to be a useful technological application. When the volume experiences significant expansion and contraction throughout the electrochemical process, bi-continuous transport of electrons and ions takes place. They successfully synthesized a coaxial semi-hollow MnO_2/porous carbon material by creating a void between the carbon material and the metal oxide nanowire (one-dimensional yolk-shell structure). A hollow structure aids in the infiltration of the electrolyte into the active material and promotes ion transport, whereas carbon facilitates quick electron conduction. In the present case, the interface charge transfer resistance is reduced from 263 to 89 Ω. At a high current density of 1 A/g^{-1}, the electrode rate exhibits a significant improvement in performance, reaching 179 mAh g^{-1}. In contrast, pure MnO_2 has an almost nonexistent capacity. Moreover, this arrangement enhances the capacity to manage surges in volume, thereby effectively hindering seismic waves. Specifically, under conditions of rapid charging and discharging, the stability of the electrode cycle will be improved throughout the cycling process to avert membrane rupture. There is a significant improvement found at a current density of 200 mA g^{-1}. Additionally, following a process of 200 cycles, the synthesized sample exhibits a specific capacity of 509 mAh g^{-1}, surpassing the 61 mAh g^{-1} capacity of pure MnO_2.

2.4.4.2 Nanostructure Array

The growth process parameters generally dictate the establishment of nanowire arrays; during this procedure, the constituent material of the nanowire array is cultivated in a consistent orientation across the substrate surface (metal-based or carbon-based material). Furthermore, apart from eliminating the need for a binder, the produced material also mitigates the electrochemical inertness binder characteristic. On the contrary to the process by which the active material adheres to the current collector, the utilization of the nanowire array presents the following advantages: A nanowire array material possessing an increased specific surface area will effectively impede the process of active material aggregation. As electrons conduct uniformly between the nanowire array and the substrate, the total conductivity of the material is significantly enhanced. The space between the nanowires places the active material in full proximity to the electrolyte, hence enabling efficient ion transport throughout the process [93].

Leveraging the benefits of nanowire arrays, Mai group examined the construction of SnO_2 polyaniline nanowire arrays that possessed high-efficiency electron transport properties [93]. A method was proposed to produce hydrothermally treated and electrodeposition-deposited heterogeneous multi-branched core–shell SnO_2-PANI nanowire arrays in a straightforward and manipulable fashion. The contrary to the structural deterioration observed in the initial SnO_2 arrays and recycled SnO_2 arrays, the complete mechanical structure of the SnO_2-PANI array, which features a multi-branched core–shell configuration, remains intact even after cycling. A robust adhesion is seen between the conductive polymer shell and the heterogeneous SnO_2-PANI array, owing to the array's unique composition and structure, which features a multi-branched core–shell arrangement. The mechanical integrity and structural stability of this array are outstanding. Concurrently, it derives advantages from the functionalities of the nanowire array, which enables the fluid advancement of three-dimensional electron and ion transport. Consequently, the rate and cycle performance of the SnO_2-PANI array are significantly improved when used as an electrode material in lithium-ion batteries. The effective implementation of the material synthesis design technique and production of this multi-branched core–shell nanowire array is likely to be applied to the creation of more high-performance energy storage devices.

2.4.5 Structural Stability Regulation and Control

Nanowire electrode materials will experience fluctuating levels of expansion or contraction throughout the electrochemical reaction, culminating in structural failure and a deterioration in operational efficacy. Sustaining the structural integrity of the nanowire electrode materials during the electrochemical reaction is thus essential for optimizing the material's performance. This section will center on the control mechanisms that are employed to guarantee the structural stability of electrode materials composed of nanowires. The internal stress that arises from the volumetric alteration of the nanowire electrode material throughout the electrochemical reaction is a significant determinant in material degradation. Internal tension caused by the reaction process can be effectively alleviated with the deployment of hierarchical and hybrid nanowire designs. Volume expansion is a structural deformation of electrode materials that transpires due to their participation in electrochemical reactions throughout the charging and discharging procedures. The identification of electrode materials that satisfy the criterion for sustained stability is an essential stride in the direction of achieving energy storage on a wide scale. The volume expansion effect can be effectively hindered by nanowires, which are one-dimensional materials characterized by their relatively stable architectures, through the establishment of a dense coating structure and cavity structure. By doing so, stress-induced electrode deterioration is substantially reduced, hence improving the stability of electrode cycles [94].

An extensive layer of supplementary materials (such as carbon, metal, or oxide) with remarkable mechanical characteristics is applied to the nanowires. Through the mechanism of volume expansion buffering, this coating protects the nanowire electrode materials. A protective barrier prevents excessive contact between the electrolyte and active substances, thereby preventing the continuous exposure of active substances and the continuous formation of solid electrolyte interface (SEI) film during volume expansion. The function of the buffer is to preserve the structure and morphology of the solution. The distinction between a homogeneous and a hetero-geneous coating layer is maintained. A coating layer that comprises nanowires in an alternative crystal phase, although possessing an identical composition, is referred to be heterogeneous. The heterogeneous coating has been the subject of extensive investigation (i.e., nanowire materials composed of various substances). A portion of the volume expansion of V_2O_5 nanowires caused by the graphene layer is attenu-ated during the insertion and extraction of Mg^{2+} ions [95]. Its exceptional cycle and magnification performance demonstrate exceptional material stability.

The implementation of close coating can efficiently prevent the increase of volume and provide structural stability. On the contrary, allowing materials some degree of flexibility to expand can also effectively contribute to the stability of the construction. Effective stability of the material structure can be achieved through the construction of cavity structures, such as a semi-hollow "pipe centerline" and a nanotube with a coating on its surface. Cavity structure nanowires consist of semi-hollow nanowires and tubular nanowires. The former possesses a radially directed inward expansion space, while the later possesses a radially directed outward expansion space. Both nanowires have the ability to create space for material expansion. Simultaneously, both possess shell protection, which enables them to provide structural stability. The cathode material for LIBs was a semihollow bi-continuous structure manufactured by Mai group using the nanowire (V_3O_7) template approach [79]. The interior void area created by the nanowire-templated graphene scroll nanostructure facilitates swelling during lithiation, while the internal electrolyte channel effectively increases the ion diffusion coefficient. A pure nanowire configuration would impede the complete and timely release of strain.

2.5 Future Directions

Practical uses of electrochemical energy storage expose nanowire electrode mate-rials to the risk of structural deterioration and conductivity degradation. This chapter is dedicated mostly to the regulation of electrical transport performance and struc-tural stability in order to improve the functionality of electrode materials made of nanowires. The regulation of ion diffusion channels, band structures, surfaces, and ion/electron dual continuous processes constitutes the majority of the burdensome control over electrical transport performance. The energy band structure can be regu-lated via the execution of the ion pre-intercalation approach, which is facilitated by the induction of an electric field that provides additional carriers and charges. Furthermore, by utilizing composite conductive materials to increase the quantity of active sites, the interface and surface were manipulated. In addition, by employing organic molecules, inorganic molecules/nonmetal ions, and metal ions during the pre-intercalation process, it is also possible to regulate the diffusion channels of ions. Moreover, by the careful fabrication of coaxial semi-hollow structures and nanowire arrays, it is feasible to achieve the simultaneous and uninterrupted management of electrons and ions.

Suppression of volume expansion and internal stress buffering are the primary components of structural stability control. The mitigation of internal stress is achieved through the use of hierarchical and hybrid nanowire architectures. Volumetric expansion can be suppressed by constructing cavities and tightly wrapped frameworks. After optimizing electrical transport efficiency and guaranteeing structural integrity, the following two considerations are crucial: (1) The integration of in situ characterization techniques with performance control is imperative for the development of high-performance nanowire electrode materials. Further focused investigation is required to ascertain the precise mechanism through which the performance of specific materials deteriorates. (2) To further the progress of micro-nano energy storage devices, secondary batteries, and supercapacitors, further research into innovative strategies and concepts that synergistically improve the electrical transport performance and structural stability of nanostructures of electrode materials is required.

References

1. K. Zhang, Z. Cen, F. Yang, K. Xu, Rational construction of $NiCo_2O_4$@Fe_2O_3 core-shell nanowire arrays for high-performance supercapacitors. Prog. Nat. Sci.: Mater. Int. **31**(1), 19–24 (2021)
2. J. Tang, Y. Ge, J. Shen, M. Ye, Facile synthesis of $CuCo_2S_4$ as a novel electrode material for ultrahigh supercapacitor performance. Chem. Commun. **52**(7), 1509–1512 (2016)
3. L. Liu, Nano-aggregates of cobalt nickel oxysulfide as a high-performance electrode material for supercapacitors. Nanoscale **5**(23), 11615–11619 (2013)
4. H. Wan, L. Li, J. Zhang, X. Liu, H. Wang, H. Wang, Nickel nanowire@porous $NiCo_2O_4$ nanorods arrays grown on nickel foam as efficient pseudocapacitor electrode. Front. Energy Res. **5**(33), 1–7 (2017)
5. Y. Miao, X. Zhang, J. Zhan, Y. Sui, J. Qi, F. Wei, Q. Meng, Y. He, Y. Ren, Z. Zhan, Z. Sun, Hierarchical NiS@CoS with controllable core-shell structure by two-step strategy for supercapacitor electrodes. Adv. Mater. Interfaces **7**(3), 1901618 (2020)
6. L. Wan, C. He, D. Chen, J. Liu, Y. Zhang, C. Du, M. Xie, J. Chen, In situ grown NiFeP@$NiCo_2S_4$ nanosheet arrays on carbon cloth for asymmetric supercapacitors. Chem. Eng. J. **399**, 125778 (2020)
7. P. Lu, X. Jiang, W. Guo, L. Wang, T. Zhang, Y. Boyjoo, W. Si, F. Hou, J. Liu, S.X. Dou, J. Liang, A Ni–Co sulfide nanosheet/carbon nanotube hybrid film for high-energy and high-power flexible supercapacitors. Carbon **178**, 355–362 (2021)
8. S. Guan, X. Fu, Z. Lao, C. Jin, Z. Peng, NiS–MoS_2 hetero-nanosheet arrays on carbon cloth for high-performance flexible hybrid energy storage devices. ACS Sustain. Chem. Eng. **7**(13), 11672–11681 (2019)
9. Y. Yan, B. Li, W. Guo, H. Pang, H. Xue, Vanadium based materials as electrode materials for high performance supercapacitors. J. Power. Sources **329**, 148–169 (2016)
10. H. Qin, S. Liang, L. Chen, Y. Li, Z. Luo, S. Chen, Recent advances in vanadium-based nano-materials and their composites for supercapacitors. Sustain. Energy Fuels **4**(10), 4902–4933 (2020)
11. W. Sun, G. Gao, Y. Du, K. Zhang, G. Wu, A facile strategy for fabricating hierarchical nanocomposites of V_2O_5 nanowire arrays on a three-dimensional N-doped graphene aerogel with a synergistic effect for supercapacitors. J. Mater. Chem. A **6**(21), 9938–9947 (2018)
12. J. Balamurugan, G. Karthikeyan, T.D. Thanh, N.H. Kim, J.H. Lee, Facile synthesis of vanadium nitride/nitrogen-doped graphene composite as stable high performance anode materials for supercapacitors. J. Power. Sources **308**, 149–157 (2016)

13. X. Wang, B. Shi, X. Wang, J. Gao, C. Zhang, Z. Yang, H. Xie, One-step synthesis of V_2O_5/Ni_3S_2 nanoflakes for high electrochemical performance. J. Mater. Chem. A **5**(45), 23543–23549 (2017)

14. I. Shakir, Z. Ali, J. Bae, J. Park, D.J. Kang, Layer by layer assembly of ultrathin V_2O_5 anchored MWCNTs and graphene on textile fabrics for fabrication of high energy density flexible supercapacitor electrodes. Nanoscale **6**(8), 4125–4130 (2014)

15. D.P. Dubal, O. Ayyad, V. Ruiz, P. Gómez-Romero, Hybrid energy storage: the merging of battery and supercapacitor chemistries. Chem. Soc. Rev. **44**(7), 1777–1790 (2015)

16. M. Mustaqeem, G.A. Naikoo, M. Yarmohammadi, M.Z. Pedram, H. Pourfarzad, R.A. Dar, S.A. Taha, I.U. Hassan, M.Y. Bhat, Y.-F. Chen, Rational design of metal oxide based electrode materials for high performance supercapacitors—a review. J. Energy Storage **55**, 105419 (2022)

17. W. Liu, M.-S. Song, B. Kong, Y. Cui, Flexible and stretchable energy storage: recent advances and future perspectives. Adv. Mater. **29**(1), 1603436 (2017)

18. H. Liu, W. Yang, Ultralong single crystalline V_2O_5 nanowire/graphene composite fabricated by a facile green approach and its lithium storage behavior. Energy Environ. Sci. **4**(10), 4000–4008 (2011)

19. W. Yuan, B. Wang, H. Wu, M. Xiang, Q. Wang, H. Liu, Y. Zhang, H. Liu, S. Dou, A flexible 3D nitrogen-doped carbon foam@CNTs hybrid hosting TiO_2 nanoparticles as free-standing electrode for ultra-long cycling lithium-ion batteries. J. Power. Sources **379**, 10–19 (2018)

20. J. Liu, L. Dong, D. Chen, Y. Han, Y. Liang, M. Yang, J. Han, C. Yang, W. He, Metal oxides with distinctive valence states in an electron-rich matrix enable stable high-capacity anodes for Li Ion batteries. Small Methods **4**(2), 1900753 (2020)

21. W. Xia, F. Xu, C. Zhu, H.L. Xin, Q. Xu, P. Sun, L. Sun, Probing microstructure and phase evolution of α-MoO_3 nanobelts for sodium-ion batteries by in situ transmission electron microscopy. Nano Energy **27**, 447–456 (2016)

22. J.S. Chen, X.W. Lou, SnO_2-based nanomaterials: synthesis and application in lithium-Ion batteries. Small **9**(11), 1877–1893 (2013)

23. Y. Deng, C. Fang, G. Chen, The developments of SnO_2/graphene nanocomposites as anode materials for high performance lithium ion batteries: a review. J. Power. Sources **304**, 81–101 (2016)

24. Z. Li, G. Wu, S. Deng, S. Wang, Y. Wang, J. Zhou, S. Liu, W. Wu, M. Wu, Combination of uniform SnO_2 nanocrystals with nitrogen doped graphene for high-performance lithium-ion batteries anode. Chem. Eng. J. **283**, 1435–1442 (2016)

25. X. Liu, T. Xu, Y. Li, Z. Zang, X. Peng, H. Wei, W. Zha, F. Wang, Enhanced X-ray photon response in solution-synthesized $CsPbBr_3$ nanoparticles wrapped by reduced graphene oxide. Sol. Energy Mater. Sol. Cells **187**, 249–254 (2018)

26. F. Ma, A. Yuan, J. Xu, P. Hu, Porous α-MoO_3/MWCNT nanocomposite synthesized via a surfactant-assisted solvothermal route as a lithium-ion-battery high-capacity anode material with excellent rate capability and cyclability. ACS Appl. Mater. Interfaces **7**(28), 15531–15541 (2015)

27. F. Lanlan, L. Zhenhuan, D. Nanping, Recent advances in vanadium-based materials for aqueous metal ion batteries: design of morphology and crystal structure, evolution of mechanisms and electrochemical performance. Energy Storage Mater. **41**, 152–182 (2021)

28. C. Jing, X.D. Liu, K. Li, X. Liu, B. Dong, F. Dong, Y. Zhang, The pseudocapacitance mechanism of graphene/CoAl LDH and its derivatives: are all the modifications beneficial? J. Energy Chem. **52**, 218–227 (2021)

29. C. Jing, B. Dong, Y. Zhang, Chemical modifications of layered double hydroxides in the supercapacitor. Energy Environ. Mater. **3**(3), 346–379 (2020)

30. X. Li, D. Du, Y. Zhang, W. Xing, Q. Xue, Z. Yan, Layered double hydroxides toward high-performance supercapacitors. J. Mater. Chem. A **5**(30), 15460–15485 (2017)

31. C. Jing, X. Song, K. Li, Y. Zhang, X. Liu, B. Dong, F. Dong, S. Zhao, H. Yao, Y. Zhang, Optimizing the rate capability of nickel cobalt phosphide nanowires on graphene oxide by the outer/inter-component synergistic effects. J. Mater. Chem. A **8**(4), 1697–1708 (2020)

32. D. Chen, X. Rui, Q. Zhang, H. Geng, L. Gan, W. Zhang, C. Li, S. Huang, Y. Yu, Persistent zinc-ion storage in mass-produced V_2O_5 architectures. Nano Energy **60**, 171–178 (2019)
33. S. Zhang, H. Tan, X. Rui, Y. Yu, Vanadium-based materials: next generation electrodes powering the battery revolution? Acc. Chem. Res. **53**(8), 1660–1671 (2020)
34. Y. Li, M. Chen, B. Liu, Y. Zhang, X. Liang, X. Xia, Heteroatom doping: an effective way to boost sodium Ion storage. Adv. Energy Mater. **10**(27), 2000927 (2020)
35. C. Liu, R. Massé, X. Nan, G. Cao, A promising cathode for Li-ion batteries: $Li_3V_2(PO_4)_3$. Energy Storage Mater. **4**, 15–58 (2016)
36. C. Zhao, Q. Wang, Z. Yao, J. Wang, B. Sánchez-Lengeling, F. Ding, X. Qi, Y. Lu, X. Bai, B. Li, H. Li, A. Aspuru-Guzik, X. Huang, C. Delmas, M. Wagemaker, L. Chen, Y.-S. Hu, Rational design of layered oxide materials for sodium-ion batteries. Science **370**(6517), 708–711 (2020)
37. R.K. Guduru, J.C. Icaza, A brief review on multivalent intercalation batteries with aqueous electrolytes. Nanomaterials **6**(3), 41 (2016)
38. R.C. Massé, C. Liu, Y. Li, L. Mai, G. Cao, Energy storage through intercalation reactions: electrodes for rechargeable batteries. Natl. Sci. Rev. **4**(1), 26–53 (2016)
39. Z. Wang, M. Zhang, W. Ma, J. Zhu, W. Song, Application of carbon materials in aqueous zinc ion energy storage devices. Small **17**(19), 2100219 (2021)
40. R. Li, F. Xing, T. Li, H. Zhang, J. Yan, Q. Zheng, X. Li, Intercalated polyaniline in V_2O_5 as a unique vanadium oxide bronze cathode for highly stable aqueous zinc ion battery. Energy Storage Mater. **38**, 590–598 (2021)
41. S. Chen, K. Li, K.S. Hui, J. Zhang, Regulation of lamellar structure of vanadium oxide via polyaniline intercalation for high-performance aqueous zinc-Ion battery. Adv. Func. Mater. **30**(43), 2003890 (2020)
42. W. Li, C. Han, Q. Gu, S.-L. Chou, J.-Z. Wang, H.-K. Liu, S.-X. Dou, Electron delocalization and dissolution-restraint in vanadium oxide superlattices to boost electrochemical performance of aqueous zinc-Ion batteries. Adv. Energy Mater. **10**(48), 2001852 (2020)
43. X.H. Rui, C. Li, J. Liu, T. Cheng, C.H. Chen, The $Li_3V_2(PO_4)_3$/C composites with high-rate capability prepared by a maltose-based sol–gel route. Electrochim. Acta **55**(22), 6761–6767 (2010)
44. D. Kong, X. Li, Y. Zhang, X. Hai, B. Wang, X. Qiu, Q. Song, Q.-H. Yang, L. Zhi, Encapsulating V_2O_5 into carbon nanotubes enables the synthesis of flexible high-performance lithium ion batteries. Energy Environ. Sci. **9**(3), 906–911 (2016)
45. W. Ren, Z. Zheng, C. Xu, C. Niu, Q. Wei, Q. An, K. Zhao, M. Yan, M. Qin, L. Mai, Self-sacrificed synthesis of three-dimensional $Na_3V_2(PO_4)_3$ nanofiber network for high-rate sodium–ion full batteries. Nano Energy **25**, 145–153 (2016)
46. H. Gamal, A.M. Elshahawy, S.S. Medany, M.A. Hefnawy, M.S. Shalaby, Recent advances of vanadium oxides and their derivatives in supercapacitor applications: a comprehensive review. J. Energy Storage **76**, 109788 (2024)
47. A. Roy, A. Ray, P. Sadhukhan, S. Saha, S. Das, Morphological behaviour, electronic bond formation and electrochemical performance study of V_2O_5-polyaniline composite and its application in asymmetric supercapacitor. Mater. Res. Bull. **107**, 379–390 (2018)
48. B. Balamuralitharan, I.-H. Cho, J.-S. Bak, H.-J. Kim, V_2O_5 nanorod electrode material for enhanced electrochemical properties by a facile hydrothermal method for supercapacitor applications. New J. Chem. **42**(14), 11862–11868 (2018)
49. N.M. Abd-Alghafour, N.M. Ahmed, Z. Hassan, M.A. Almessiere, Hydrothermal synthesis and structural properties of V_2O_5 nanoflowers at low temperatures. J. Phys. Conf. Ser. **1083**(1), 012036 (2018)
50. C. Peng, M. Jin, D. Han, X. Liu, L. Lai, Structural engineering of V_2O_5 nanobelts for flexible supercapacitors. Mater. Lett. **320**, 132391 (2022)
51. Z. Xue, K. Tao, L. Han, Stringing metal–organic framework-derived hollow Co_3S_4 nanopolyhedra on V_2O_5 nanowires for high-performance supercapacitors. Appl. Surf. Sci. **600**, 154076 (2022)
52. W. Bi, J. Wang, E.P. Jahrman, G.T. Seidler, G. Gao, G. Wu, G. Cao, Interface engineering V_2O_5 nanofibers for high-energy and durable supercapacitors. Small **15**(31), 1901747 (2019)

53. X. Jing, Y. Zhang, H. Jiang, Y. Cheng, N. Xing, C. Meng, Facile template-free fabrication of hierarchical V_2O_5 hollow spheres with excellent charge storage performance for symmetric and hybrid supercapacitor devices. J. Alloy. Compd. **763**, 180–191 (2018)
54. M. Li, T. Ai, L. Kou, J. Song, W. Bao, Y. Wang, X. Wei, W. Li, Z. Deng, X. Zou, H. Wang, Synthesis and electrochemical performance of V_2O_5 nanosheets for supercapacitor. AIP Adv. **12**(5), 055203 (2022)
55. G. Sun, H. Ren, Z. Shi, L. Zhang, Z. Wang, K. Zhan, Y. Yan, J. Yang, B. Zhao, V_2O_5/ vertically-aligned carbon nanotubes as negative electrode for asymmetric supercapacitor in neutral aqueous electrolyte. J. Colloid Interface Sci. **588**, 847–856 (2021)
56. M.D. Patil, S.D. Dhas, A.A. Mane, A.V. Moholkar, Clinker-like V_2O_5 nanostructures anchored on 3D Ni-foam for supercapacitor application. Mater. Sci. Semicond. Process. **133**, 105978 (2021)
57. J. Yang, T. Lan, J. Liu, Y. Song, M. Wei, Supercapacitor electrode of hollow spherical V_2O_5 with a high pseudocapacitance in aqueous solution. Electrochim. Acta **105**, 489–495 (2013)
58. J. Zheng, Y. Zhang, T. Hu, T. Lv, C. Meng, New strategy for the morphology-controlled synthesis of V_2O_5 microcrystals with enhanced capacitance as battery-type supercapacitor electrodes. Cryst. Growth Des. **18**(9), 5365–5376 (2018)
59. P. He, G. Zhang, X. Liao, M. Yan, X. Xu, Q. An, J. Liu, L. Mai, Sodium Ion stabilized vanadium oxide nanowire cathode for high-performance zinc-Ion batteries. Adv. Energy Mater. **8**(10), 1702463 (2018)
60. Q. Wei, Z. Jiang, S. Tan, Q. Li, L. Huang, M. Yan, L. Zhou, Q. An, L. Mai, Lattice breathing inhibited layered vanadium oxide ultrathin nanobelts for enhanced sodium storage. ACS Appl. Mater. Interfaces **7**(33), 18211–18217 (2015)
61. Y. Liu, T. Zhou, Y. Zheng, Z. He, C. Xiao, W.K. Pang, W. Tong, Y. Zou, B. Pan, Z. Guo, Y. Xie, Local electric field facilitates high-performance Li-Ion batteries. ACS Nano **11**(8), 8519–8526 (2017)
62. J. Jeong, N. Aetukuri, T. Graf, T.D. Schladt, M.G. Samant, S.S.P. Parkin, Suppression of metal-insulator transition in VO_2 by electric field–induced oxygen vacancy formation. Science **339**(6126), 1402–1405 (2013)
63. T. Koketsu, J. Ma, B.J. Morgan, M. Body, C. Legein, W. Dachraoui, M. Giannini, A. Demortière, M. Salanne, F. Dardoize, H. Groult, O.J. Borkiewicz, K.W. Chapman, P. Strasser, D. Dambournet, Reversible magnesium and aluminium ions insertion in cation-deficient anatase TiO_2. Nat. Mater. **16**(11), 1142–1148 (2017)
64. K.J. Takeuchi, S.Z. Yau, M.C. Menard, A.C. Marschilok, E.S. Takeuchi, Synthetic control of composition and crystallite size of silver hollandite, $Ag_xMn_8O_{16}$: impact on electrochemistry. ACS Appl. Mater. Interfaces **4**(10), 5547–5554 (2012)
65. S. Lian, C. Sun, W. Xu, W. Huo, Y. Luo, K. Zhao, G. Yao, W. Xu, Y. Zhang, Z. Li, K. Yu, H. Zhao, H. Cheng, J. Zhang, L. Mai, Built-in oriented electric field facilitating durable $ZnMnO_2$ battery. Nano Energy **62**, 79–84 (2019)
66. M. Yan, X. Pan, P. Wang, F. Chen, L. He, G. Jiang, J. Wang, J.Z. Liu, X. Xu, X. Liao, J. Yang, L. Mai, Field-effect tuned adsorption dynamics of VSe_2 nanosheets for enhanced hydrogen evolution reaction. Nano Lett. **17**(7), 4109–4115 (2017)
67. X. Liu, Z. Li, X. Liao, X. Hong, Y. Li, C. Zhou, Y. Zhao, X. Xu, L. Mai, A three-dimensional nitrogen-doped graphene framework decorated with an atomic layer deposited ultrathin V_2O_5 layer for lithium sulfur batteries with high sulfur loading. J. Mater. Chem. A **8**(24), 12106–12113 (2020)
68. J.-S. Park, S. Yang, Y.C. Kang, Boosting the electrochemical performance of V_2O_3 by anchoring on carbon nanotube microspheres with macrovoids for ultrafast and long-life aqueous zinc-Ion batteries. Small Methods **5**(9), 2100578 (2021)
69. S. Deng, Z. Yuan, Z. Tie, C. Wang, L. Song, Z. Niu, Electrochemically induced metal–organic-framework-derived amorphous V_2O_5 for superior rate aqueous zinc-Ion batteries. Angew. Chem. Int. Ed. **59**(49), 22002–22006 (2020)
70. Z. Zhang, B. Xi, X. Wang, X. Ma, W. Chen, J. Feng, S. Xiong, Oxygen defects engineering of $VO_2 \cdot xH_2O$ nanosheets via in situ polypyrrole polymerization for efficient aqueous zinc-Ion storage. Adv. Funct. Mater. **31**(34), 2103070 (2021)

71. Z. Wang, X. Tang, S. Yuan, M. Bai, H. Wang, S. Liu, M. Zhang, Y. Ma, Engineering vanadium pentoxide cathode for the zero-strain cation storage via a scalable intercalation-polymerization approach. Adv. Func. Mater. **31**(22), 2100164 (2021)

72. Q. Wei, Q. An, D. Chen, L. Mai, S. Chen, Y. Zhao, K.M. Hercule, L. Xu, A. Minhas-Khan, Q. Zhang, One-pot synthesized bicontinuous hierarchical $Li_3V_2(PO_4)_3$/C mesoporous nanowires for high-rate and ultralong-life lithium-ion batteries. Nano Lett. **14**(2), 1042–1048 (2014)

73. X. Xiao, X. Peng, H. Jin, T. Li, C. Zhang, B. Gao, B. Hu, K. Huo, J. Zhou, Freestanding mesoporous VN/CNT hybrid electrodes for flexible all-solid-state supercapacitors. Adv. Mater. **25**(36), 5091–5097 (2013)

74. Y. Dai, X. Liao, R. Yu, J. Li, J. Li, S. Tan, P. He, Q. An, Q. Wei, L. Chen, X. Hong, K. Zhao, Y. Ren, J. Wu, Y. Zhao, L. Mai, Quicker and more Zn^{2+} storage predominantly from the interface. Adv. Mater. **33**(26), 2100359 (2021)

75. Y. Dai, Q. Li, S. Tan, Q. Wei, Y. Pan, X. Tian, K. Zhao, X. Xu, Q. An, L. Mai, Q. Zhang, Nanoribbons and nanoscrolls intertwined three-dimensional vanadium oxide hydrogels for high-rate lithium storage at high mass loading level. Nano Energy **40**, 73–81 (2017)

76. X. Tian, X. Xu, L. He, Q. Wei, M. Yan, L. Xu, Y. Zhao, C. Yang, L. Mai, Ultrathin pre-lithiated V_6O_{13} nanosheet cathodes with enhanced electrical transport and cyclability. J. Power. Sources **255**, 235–241 (2014)

77. Q. An, P. Zhang, Q. Wei, L. He, F. Xiong, J. Sheng, Q. Wang, L. Mai, Top-down fabrication of three-dimensional porous V_2O_5 hierarchical microplates with tunable porosity for improved lithium battery performance. J. Mater. Chem. A **2**(10), 3297–3302 (2014)

78. Y. Cheng, J. Shu, L. Xu, Y. Xia, L. Du, G. Zhang, L. Mai, Flexible nanowire cathode membrane with gradient interfaces and rapid electron/Ion transport channels for solid-state lithium batteries. Adv. Energy Mater. **11**(12), 2100026 (2021)

79. M. Yan, F. Wang, C. Han, X. Ma, X. Xu, Q. An, L. Xu, C. Niu, Y. Zhao, X. Tian, P. Hu, H. Wu, L. Mai, Nanowire templated semihollow bicontinuous graphene scrolls: designed construction, mechanism, and enhanced energy storage performance. J. Am. Chem. Soc. **135**(48), 18176–18182 (2013)

80. Z. Cai, L. Xu, M. Yan, C. Han, L. He, K.M. Hercule, C. Niu, Z. Yuan, W. Xu, L. Qu, K. Zhao, L. Mai, Manganese oxide/carbon yolk-shell nanorod anodes for high capacity lithium batteries. Nano Lett. **15**(1), 738–744 (2015)

81. J.G. Kim, S.H. Nam, S.H. Lee, S.M. Choi, W.B. Kim, SnO_2 nanorod-planted graphite: an effective nanostructure configuration for reversible lithium Ion storage. ACS Appl. Mater. Interfaces **3**(3), 828–835 (2011)

82. K. Kang, G. Ceder, Factors that affect Li mobility in layered lithium transition metal oxides. Phys. Rev. B **74**(9), 094105 (2006)

83. V. Petkov, P.N. Trikalitis, E.S. Bozin, S.J.L. Billinge, T. Vogt, M.G. Kanatzidis, Structure of $V_2O_5 \cdot nH_2O$ xerogel solved by the atomic pair distribution function technique. J. Am. Chem. Soc. **124**(34), 10157–10162 (2002)

84. Y.-N. Zhou, J. Ma, E. Hu, X. Yu, L. Gu, K.-W. Nam, L. Chen, Z. Wang, X.-Q. Yang, Tuning charge–discharge induced unit cell breathing in layer-structured cathode materials for lithium-ion batteries. Nat. Commun. **5**(1), 5381 (2014)

85. Y. Zhao, C. Han, J. Yang, J. Su, X. Xu, S. Li, L. Xu, R. Fang, H. Jiang, X. Zou, B. Song, L. Mai, Q. Zhang, Stable alkali metal ion intercalation compounds as optimized metal oxide nanowire cathodes for lithium batteries. Nano Lett. **15**(3), 2180–2185 (2015)

86. L. Peng, Y. Zhu, X. Peng, Z. Fang, W. Chu, Y. Wang, Y. Xie, Y. Li, J.J. Cha, G. Yu, Effective interlayer engineering of two-dimensional $VOPO_4$ nanosheets via controlled organic intercalation for improving alkali Ion storage. Nano Lett. **17**(10), 6273–6279 (2017)

87. M. Clites, E. Pomerantseva, Bilayered vanadium oxides by chemical pre-intercalation of alkali and alkali-earth ions as battery electrodes. Energy Storage Mater. **11**, 30–37 (2018)

88. M. Yan, P. He, Y. Chen, S. Wang, Q. Wei, K. Zhao, X. Xu, Q. An, Y. Shuang, Y. Shao, K.T. Mueller, L. Mai, J. Liu, J. Yang, Water-lubricated intercalation in $V_2O_5 \cdot nH_2O$ for high-capacity and high-rate aqueous rechargeable zinc batteries. Adv. Mater. **30**(1), 1703725 (2018)

89. M. Clites, J.L. Hart, M.L. Taheri, E. Pomerantseva, Chemically preintercalated bilayered $K_xV_2O_5 \cdot nH_2O$ nanobelts as a high-performing cathode material for K-Ion batteries. ACS Energy Lett. **3**(3), 562–567 (2018)

90. Y. Xu, X. Deng, Q. Li, G. Zhang, F. Xiong, S. Tan, Q. Wei, J. Lu, J. Li, Q. An, L. Mai, Vanadium oxide pillared by interlayer Mg^{2+} ions and water as ultralong-life cathodes for magnesium-Ion batteries. Chem **5**(5), 1194–1209 (2019)

91. L. Zhou, Q. Liu, Z. Zhang, K. Zhang, F. Xiong, S. Tan, Q. An, Y.-M. Kang, Z. Zhou, L. Mai, Interlayer-spacing-regulated $VOPO_4$ nanosheets with fast kinetics for high-capacity and durable rechargeable magnesium batteries. Adv. Mater. **30**(32), 1801984 (2018)

92. J. Dong, Y. Jiang, Q. Wei, S. Tan, Y. Xu, G. Zhang, X. Liao, W. Yang, Q. Li, Q. An, L. Mai, Strongly coupled pyridine-$V_2O_5 \cdot nH_2O$ nanowires with intercalation pseudocapacitance and stabilized layer for high energy sodium ion capacitors. Small **15**(22), 1900379 (2019)

93. W. Xu, K. Zhao, C. Niu, L. Zhang, Z. Cai, C. Han, L. He, T. Shen, M. Yan, L. Qu, L. Mai, Heterogeneous branched core–shell SnO_2–PANI nanorod arrays with mechanical integrity and three dimentional electron transport for lithium batteries. Nano Energy **8**, 196–204 (2014)

94. W. Zhou, C. Cheng, J. Liu, Y.Y. Tay, J. Jiang, X. Jia, J. Zhang, H. Gong, H.H. Hng, T. Yu, H.J. Fan, Epitaxial growth of branched α-Fe_2O_3/SnO_2 nano-heterostructures with improved lithium-Ion battery performance. Adv. Func. Mater. **21**(13), 2439–2445 (2011)

95. Q. An, Y. Li, H. Deog Yoo, S. Chen, Q. Ru, L. Mai, Y. Yao, Graphene decorated vanadium oxide nanowire aerogel for long-cycle-life magnesium battery cathodes. Nano Energy **18**, 265–272 (2015)

Chapter 3
Vanadium-Based Cathodes for Supercapacitors (Adapted with Elsevier Copyright 2022, Order Number: 5842360816511)

3.1 Implementing Electrodeposition Method for Cathode Materials

The electrodeposition method is one of the most convenient methods for synthesizing cathode materials for supercapacitors. The electrolyte can be easily switched from one precursor to another for material modification during the synthesis process. In a typical case of electrodeposition to fabricate V-based cathodes, vanadium precursor of VCl_3 is used as the main source with different transition metal ions as the dopants. VCl_3 is used as the matrix and mixed with other metal precursors (KCl, $LiCl$, $NiCl_2$) with a concentration ratio of 1: 0.5 (0.05 M: 0.025 M) in 50 mL deionized water for the electrolyte. Ni foam, Pt plate, and Ag/AgCl/KCl are used as the working, counter, and reference electrodes during the electrodeposition process in a workstation of SP-300. A constant current was applied with a fixed current of -30 mA for three minutes. After electroplating, the sample is soaked in deionized water to remove excess plating solution. The sample is then placed in an oven and dried at 60 °C overnight to obtain the sample. The vanadium-based films prepared from electroplating solutions with different components are denoted as VK, VL, and VN. After electrochemical testing, the VN with the best performance is selected, and subsequent optimization of experimental parameters is adjusted.

Based on the obtained data in Fig. 3.1 using Eq. (1.3), the specific capacitances of V-based cathode with different dopants can be calculated and the results are shown in Table 3.1. The results indicate that Ni dopant induces higher capacitance of V-based cathode.

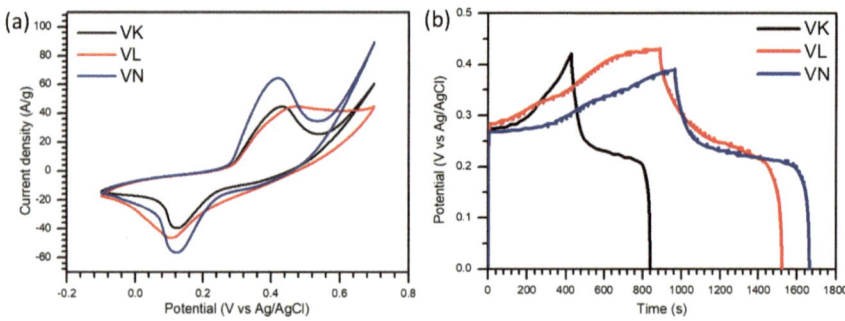

Fig. 3.1 a CV and **b** GCD spectra of V-based films on Ni foam with different dopants of potassium, lithium, and nickel precursors

Table 3.1 Specific capacitances of VK, VL, and VN cathodes

Sample	Specific capacitance (F/g at 1 A/g)
VK	980
VL	1538
VN	1994

3.2   Adjusting the Electrodeposition Conditions to Improve the Capacitance of V-Based Cathodes

Electrolyte composition during the electrodeposition process is crucial to modify the surface chemical states. Consistently, the VN electrode capacitance can be improved by changing the precursor concentrations. In addition, the reaction time and current density are also important to be adjusted to obtain the highest capacitance of the as-designed electrode. To optimize the electrochemical performances, the experimental designs of VN cathodes are done with different precursor concentration ratios, electrodeposition times, and current densities, as shown in Tables 3.2, 3.3, and 3.4.

Table 3.2 *V*-based cathodes synthesized with different precursor concentration ratios

Cathodes VNx ($x = 0, 0.25, 0.5, 1, 2$, and 4)	V and Ni precursor concentration ratios (VCl$_3$: NiCl$_2$)
VN0	1:0
VN0.25	1:0.25
VN0.5	1:0.5
VN1	1:1
VN2	1:2
VN4	1:4

Table 3.3 V-based cathodes synthesized with different current densities

Cathodes VN1y ($y = A, B, C, D$)	Current densities (mA)
VN1A	−5
VN1B	−10
VN1C	−20
VN1D	−30

Table 3.4 V-based cathodes synthesized with different electrodeposition times

Cathodes VN1Bz ($z = 1, 2, 3, 4, 5, 7$)	Electrodeposition times (min)
VN1B1	1
VN1B2	2
VN1B3	3
VN1B4	4
VN1B5	5
VN1B7	7

3.2.1 XRD and Raman Analysis of VN1 Prepared with Different Electrolyte Compositions

V-based cathodes with different precursor concentration ratios are examined with XRD measurements and the results are shown in Fig. 3.2. An X-ray diffractometer was used to assess the XRD patterns of electrodes that were manufactured with varying concentration ratios. Figure 3.2a exhibits XRD patterns of as-deposited VNx cathodes with different Ni concentrations. XRD patterns showed that all of the cathode materials that were deposited on Ni foam had the same crystal structure, even though the electrodeposition was carried out with different concentrations of $NiCl_2$. The crystal structure of VN during the electrodeposition process was unaffected by the addition of Ni precursor to the electrolyte. Figure 3.2b shows the XRD patterns of VN1, standard PDF #71–0039 of V_3O_5, and Ni foam substrate. During a 5-min electrodeposition operation, the VN1 electrode, a typical VN sample, was created with a current density of 10 mA/cm^2. The diffraction peaks of VN1 and V_3O_5 complemented each other quite well. Furthermore, the (110) peak of VN-1 was changed from V_3O_5 at 22.0° to a higher angle at 22.4°. With the greater ionic radius of Ni^{2+} than that of V^{3+}/V^{4+} in VN-1, the lattice expansion caused by the Ni addition may have played a role in the lattice sites. Ni-doped V_3O_5 phase for the Ni-foam electrode was validated by the XRD data. Notably, certain XRD peaks are absent, which may be explained by the charge transfer that occurs during electrodeposition or by the Ni inclusion that modifies the behavior of the crystal growth during heterogeneous nucleation.

Figure 3.3 shows the oxyvanite structure of monoclinic V_3O_5. The oxygen atoms are grouped in a hexagonally tight pack, while the vanadium atoms occupy 3/5 of the

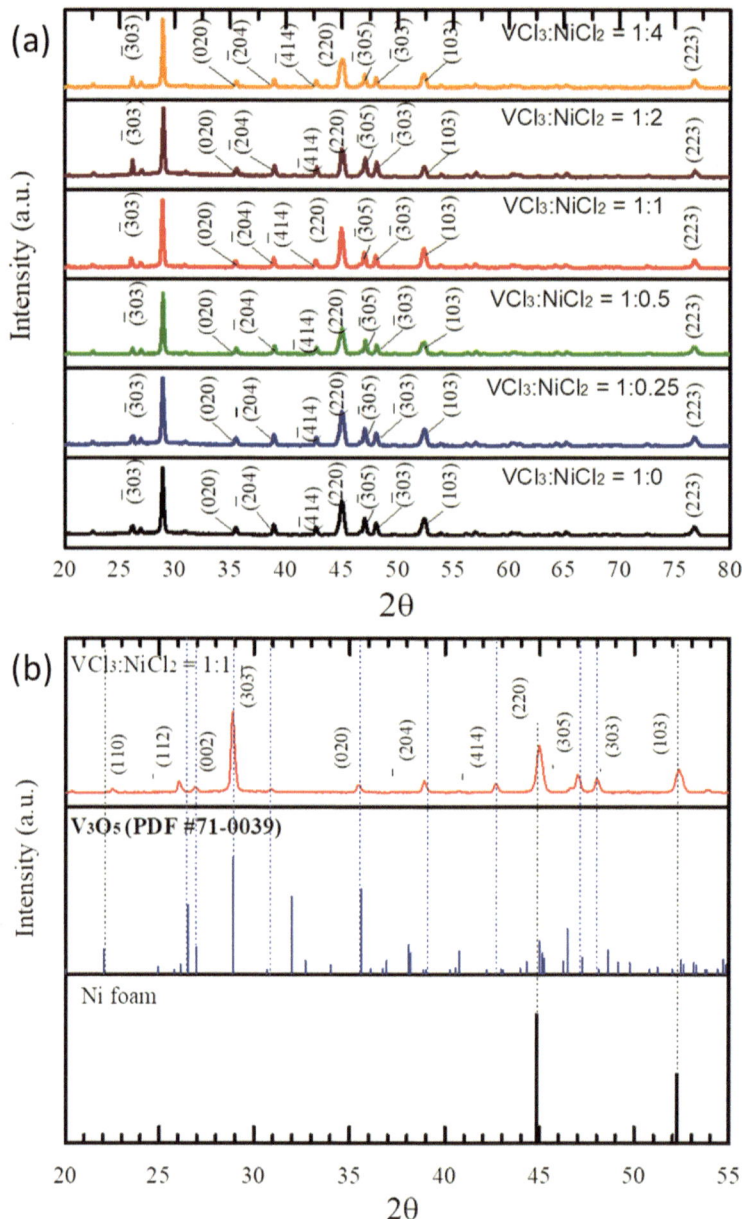

Fig. 3.2 XRD patterns of **a** VNx (x = 0, 0.25, 0.5, 1, 2, and 4) and **b** VN1 compared to the standard V₃O₅ (PDF #71–0039) and Ni-foam substrate

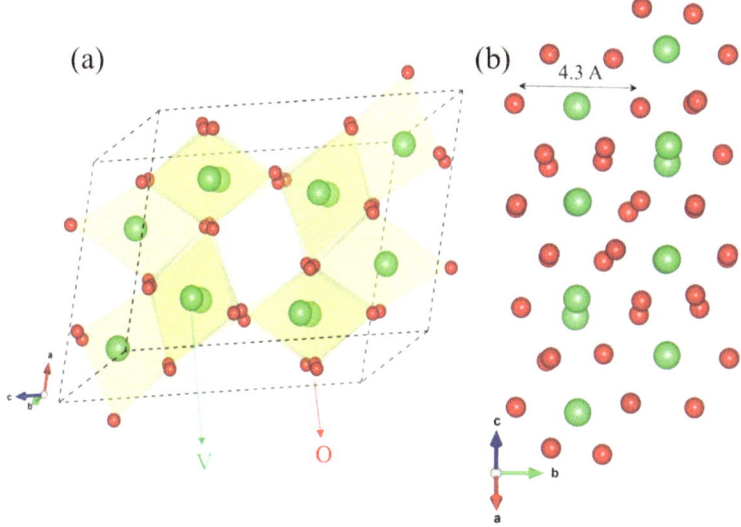

Fig. 3.3 Illustrations of the representative oxyvanite V_3O_5 crystal structure. (Elsevier Copyright 2022 with quote number: 501883039)

octahedral interstices. Consequently, four distinct forms of distorted VO_6 octahedra are produced, which together with the chain-connected open frame structure in three dimensions, share similar corners, edges, or faces to create characteristic chains (parallel to [001]). Through such a framework, potassium ions with several open regions may be accommodated in the V_3O_5 structure. Moreover, V^{3+} and V^{4+} are present in the oxyvanite system concurrently, which may help some ionic species intercalate or insert themselves into spacious lattice V_3O_5 (4.3 pm) [1]. To further evaluate the structure of VN electrodes, Raman spectra were performed, as shown in Fig. 3.4. All surface phonon vibrations of VN were very broad, which may be caused by the chemical composition, crystal size, and molecular chain length [2]. VN1 is the only one to exhibit a broad peak around 100–400 cm^{-1}, which corresponds to the range of the V_3O_5 Raman active modes of A_g and B_g.

3.2.2 SEM Analysis of VN1 Prepared with Different Electrolyte Compositions

FE-SEM and TEM investigations were used to look at the microstructure and morphology of the cathode material as it was placed over Ni foam. The SEM images of VN1 at low and high magnifications are displayed in Fig. 3.5a and 3.5b, respectively. The macroscopic morphology of porous lamellae on Ni foam was uniformly displayed in the SEM image. The size of some of the lamellae pores ranged from 50 to 100 nm. It is anticipated that the porous surface will offer a large surface

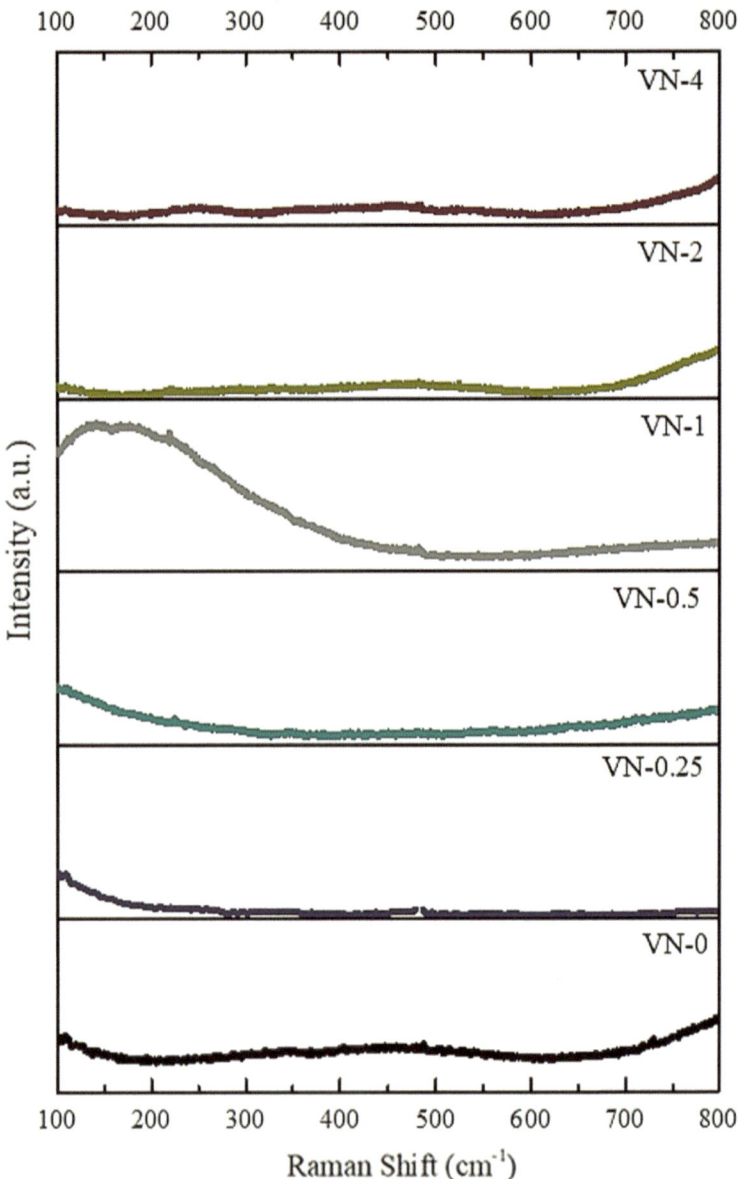

Fig. 3.4 Raman spectra of VNx cathodes with different precursor concentration ratios. (Elsevier Copyright 2022 with quote number: 501883039)

area to assist ion transfer and the intercalation reaction during the charge storage process. Furthermore, the uniformly distributed elements in VN1 with a reduced Ni concentration are visible in the SEM element mapping pictures of Ni, V, and O. Moreover, the nanostructure of VN1 was examined using the TEM method. The standing sheet-like VN1 particles on Ni foam, which can be employed for charge and discharge processes with ledges as active sites, are visible in the low magnification TEM image in Fig. 3.5c. As seen in Fig. 3.5d, some lattice fringes are visible in the high-magnification image. It has been noted that the comparable d-spacing values at $(\overline{4}\,14)$, (121), and $(\overline{2}\,04)$ were 2.10, 2.05, and 2.30 Å, respectively. The planes of V_3O_5 are represented by those discernible lattice fringes, as shown by the XRD pattern. It is also consistent that the linked planes of $(\overline{4}\,14)$, (121), and $(\overline{2}\,04)$ and (220) are suggested by the selected area electron diffraction (SAED) pattern (Fig. 3.5e). The V_3O_5 phase is confirmed by the TEM examination, which also shows that the VN1 shape and lattice fringes are consistent with the SEM and XRD investigations, respectively.

3.2.3 XPS Analysis of VN1 Prepared with Different Electrolyte Compositions

A potent XPS measurement was used to examine the surface chemical state and composition of the uppermost layer of VN1 on a Ni-foam substrate. The high-resolution spectra of V 2p, Ni 2p, and O 1 s orbitals for the fresh and used VN1 cathodes are shown in Fig. 3.6. The binding energies for the V $2p_{3/2}$ and $2p_{1/2}$ orbitals, which are positioned at 515.5 and 523.5 eV, respectively, are shown in Fig. 3.6a. These values are in line with earlier research and are associated with the V^{4+} state [3]. In addition, V $2p_{3/2}$ and V $2p_{1/2}$ orbitals for V^{3+} were also observed at 514.4 and 522.4 eV, respectively. Based on the peak area, the amount ratio of V^{4+} and V^{3+} was about 0.5, indicating the oxyvanite structure [1]. A satellite peak between those V peaks was also observed, which was identified as the O 1 s core level [4]. The oxidation states of V^{4+} and V^{3+} agree with the suggested oxyvanite phase in XRD analysis. Figure 3.6b shows the characteristic peaks of Ni 2p orbitals. The state of Ni dopant in the V_3O_5 phase was identified as Ni^{2+} with $2p_{3/2}$ and $2p_{1/2}$ orbitals observed at 855.65 and 873.18 eV, respectively [5, 6]. Furthermore, the distinctive satellite peaks of Ni 2p orbitals were displayed. The reductive electrode-position procedure generates a small quantity of Ni metal at 850.31 eV. The states of O 1 s in the VN1 electrode are shown in Fig. 3.6c. The oxygen states were identified as oxygen in the V_3O_5 lattice (O_L) and adsorbed hydroxide on the electrode surface (OH_{ads}). OH_{ads} and O_L matched with the binding energy values of 531.8 and 530.2 eV, respectively [7, 8].

Based on the peak area and sensitivity factor of each element in the VN1 cathode, a quantitative analysis of its surface composition was conducted. The elemental percentages of VN1 are shown in a table in Fig. 3.6. The XPS examination reveals

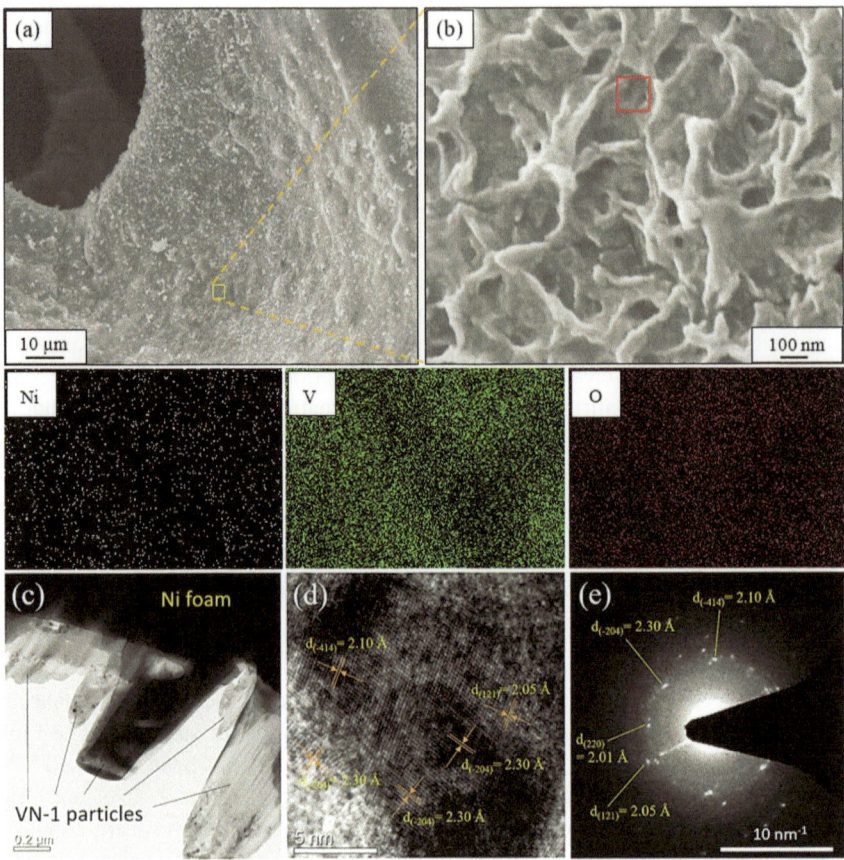

Fig. 3.5 a, b SEM images of VN1 cathode morphology with its elemental mapping and **c** its TEM image with **d** lattice fringes and **e** SAED pattern. (Elsevier Copyright 2022 with quote number: 501883039)

that V and O make up the majority, with trace levels of Ni^0 and Ni^{2+}. The as-prepared VN1 lattice had 28.7% and 54.8% vanadium and oxygen, respectively, demonstrating the consistency of the XRD study for the V_3O_5 phase. The maximum surface examination on newly synthesized VN1 showed 6.7% Ni^{2+} and 0.7% Ni^0, indicating significantly low amounts of Ni that doped in V_3O_5.

After being employed in the cycle test, the VN1 electrode was reexamined using XPS analysis to investigate the elemental chemical states. Following the 10-K cycling test, the binding energy of V in VN1 is displayed in Fig. 3.6d. The structure still contains the oxidation states of V^{4+} and V^{3+}, whose $2p_{3/2}$ and $2p_{1/2}$ orbitals are detected at 514.7 and 522.2 eV as well as 515.7 and 523.3 eV, respectively. The employed V peak locations stay the same as those of the VN1 that is depicted in Fig. 3.6a as it was prepared. However, an additional V metal with $2p_{1/2}$ orbital at 519.8 eV with a relatively small amount was observed in the spectra, implying small

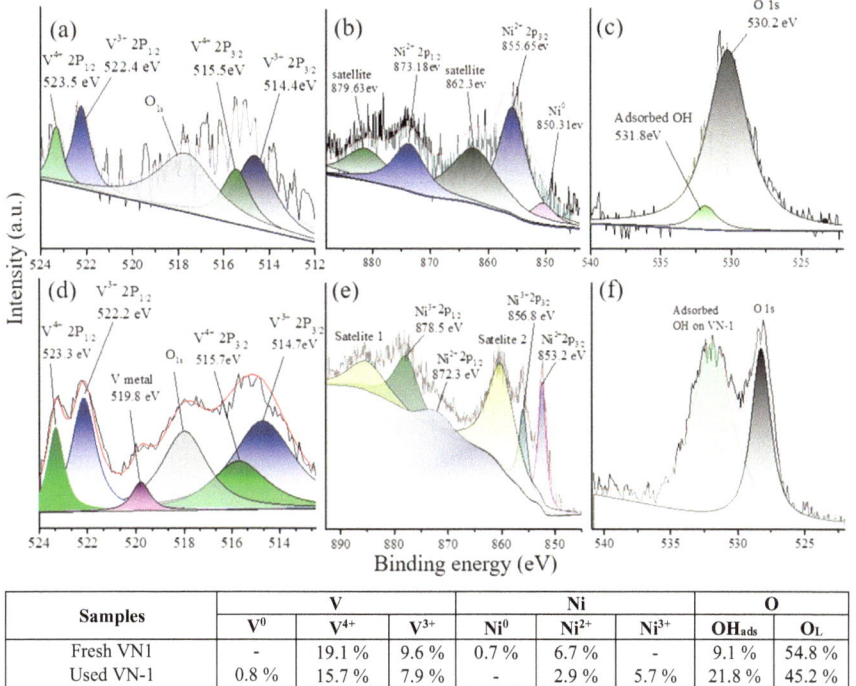

Fig. 3.6 High-resolution XPS spectra of **a** V 2p, **b** Ni 2p, and **c** O 1 s orbitals in fresh and used VN-1 cathodes with a table to describe their compositions. (Elsevier Copyright 2022 with quote number: 501883039)

Samples	V			Ni			O	
	V^0	V^{4+}	V^{3+}	Ni^0	Ni^{2+}	Ni^{3+}	OH_{ads}	O_L
Fresh VN1	-	19.1 %	9.6 %	0.7 %	6.7 %	-	9.1 %	54.8 %
Used VN-1	0.8 %	15.7 %	7.9 %	-	2.9 %	5.7 %	21.8 %	45.2 %

amounts of V-oxide were reduced to V metal during the cycling test [9]. Regarding the Ni XPS spectra, the new $2p_{3/2}$ and $2p_{1/2}$ orbitals can be seen at 856.8 and 878.5 eV, respectively, in Fig. 3.6e [10]. This indicates that the chemical state of Ni has partially shifted from Ni^{2+} to Ni^{3+}. O_{ads} and O 1 s spectra, displayed in Fig. 3.6f, are verified to be 532.0 and 528.3 eV, respectively. Fascinatingly, the pseudocapacitive OH-anion adsorption on the VN1 cathode surface causes the adsorbed oxygen to be greatly increased following the charge and discharge processes. After 10 K-cycle tests, the V content in VN1 may have decreased due to V dissolving in a basic PVA/KOH gel solution during the charge and discharge procedure.

3.2.4 Electrochemical Analysis of VN1 Prepared with Different Electrolyte Compositions

CV, GCD, and EIS measurements were used to examine the electrochemical charac-teristics of as-deposited cathodes under various synthesis circumstances to ascertain the Cs, Ed, and Pd. The CV, GCD, EIS, and Cs plots of as-deposited cathodes with varying solution concentrations are displayed in Fig. 3.7 after the electrodeposition procedure, which lasted for 5 min at a constant current density at -10 mA/cm^2. The various CV profiles of the VNx (x = 0, 0.25, 0.5, 1, 2, and 4) samples at a scan rate of 30 mV/s in the potential range of -0.1 to 0.7 V are shown in Fig. 3.7a. For every VNx sample, the plot indicates two Faradaic redox peaks, suggesting an electrode behavior similar to that of a battery. Among the six samples, VN1 delivers the largest CV integral area and the highest redox peak current, suggesting that VN1 has a greater specific capacity. The enlarged electroactive sites and electrode/electrolyte contact regions of VN1 may be the source of its enhanced redox activity [11]. The GCD curves of VNx cathodes at 1 A/g are shown in Fig. 3.7b. A battery-like characteristic is also seen in the GCD profile, which is consistent with the CV data. It is observed that compared to other VNx cathodes, the discharge period of VN1 is significantly longer. As a result, VN1 achieves greater specific capacity, energy density, and power density at 4166 F/g, 88 Wh/kg, and 186 W/kg, respectively. The EIS spectra of VNx cathodes are shown in Fig. 3.7c. An equivalent electrical circuit with equivalent series resistance (R_s), charge transfer resistance (R_{ct}), constant-phase element (CPE), and Warburg impedance is fitted to the Nyquist plots (W). In the high-frequency range, the intercept of the plots with the x-axis indicates Rs (intrinsic resistance of active materials, ionic resistance, contact resistance). Better ionic transport to improve diffusion kinetics is suggested by the linear line with a steeper slope in the low-frequency zone, which also suggests a reduced Warburg impedance. While the Warburg and R_s impedance levels were similar, their R_{ct} values were clearly different. For VN0, VN0.25, VN0.5, VN1, VN2, and VN4, the corresponding R_{ct} values were 0.39, 0.64, 0.02, 0.01, 0.47, and 0.03 Ω, respectively. According to the data, VN1 has the fastest charge transition and the best electrical conductivity. Figure 3.7d consistently displays the computed specific capacity of VNx cathodes, with the VN1 cathode displaying the highest Cs of around 4166 F/g in comparison to other VNx cathodes. According to the experimental findings, altering the composition of the V and Ni precursors is crucial for modifying the electrode surface chemistry. It consequently results in a surface electrical characteristic that is distinct in terms of conductivity and redox potentials.

Based on the best electrochemical performances of VN1, the preparation of the VN1 cathode is further improved using different conditions of electrodeposition by varying current density and reaction time, as shown in Tables 3.3 and 3.4, respectively.

The electrodeposition processes were performed with different current densities at -5, -10, -20, and -30 mA/cm^2 for 5 min to further increase the charge storage capacity of the VN1 cathode. The CV spectra with a scan rate of 30 mV/s in Fig. 3.8a demonstrate how the electrodeposition current density increases with the Faradaic-type oxidation and reduction current densities. A battery-like feature with a plateau at around 0.2 V is revealed by the GCD plots at 1 A/g in Fig. 3.8b for VN1 generated at various electrodeposition current densities. The as-prepared electrode discharge times match the CV area that was attained, as shown in Fig. 3.8a. Furthermore, the VN1 discharge time deposited at 10 mA/cm^2 grew noticeably longer, suggesting a larger energy density. VN1 exhibits improved Cs, Ed, and Pd values of 5529 F/g, 116 Wh/kg, and 204 W/Kg, respectively, by depositing the sample over Ni foam at

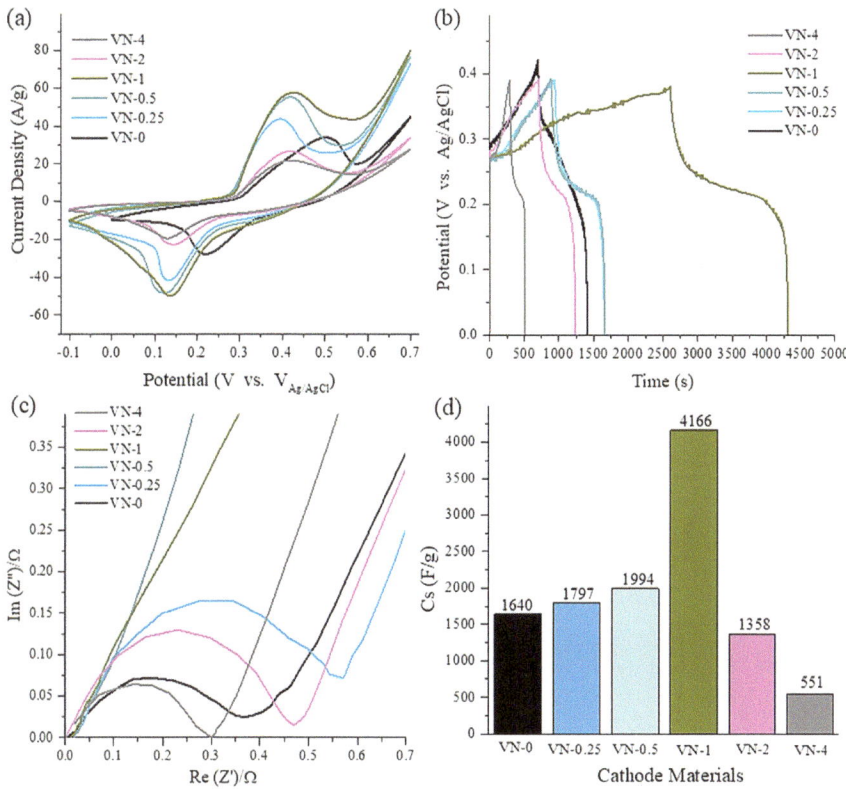

Fig. 3.7 **a** CV, **b** GCD, **c** EIS, and **d** the specific capacities of VNx cathodes for $x = 0$, 0.25, 0.5, 1, 2, and 4 synthesized at -30 mA/cm^2 electrodeposition process for 5 min. (Elsevier Copyright 2022 with quote number: 501883039)

a rate of -10 mA/cm^2. The VN1 Nyquist charts at various electrodeposition current densities of -5, -10, -20, and -30 mA/cm^2 are shown in Fig. 3.8c. There was no variation in the Warburg impedance profile across the spectra when a supercapacitor was used. On the other hand, the ionic and charge transfer resistivities changed in response to the various electrodeposition current densities. VN1 produced with a current density of -10 mA/cm^2 had the lowest values of R_s and R_{ct}. This is consistent with a lower intrinsic resistivity, which results in a longer discharging time in GCD analysis. Therefore, in comparison to other cultures, VN-1 cultivated at -10 mA/cm^2 demonstrated a greater Cs value.

The electrodeposition procedure was repeated with varying deposition periods at 1, 2, 3, 4, 5, and 7 min in order to better optimize the VN1 capacity. As illustrated in Fig. 3.9, the as-deposited VN1 cathode was investigated using CV, GCD, and EIS tests to determine their particular capabilities. With varying reaction times, the CV spectra in Fig. 3.9a, which were obtained at a scan rate of 30 mV/s, did not change noticeably, suggesting that the reaction conditions had presumably been optimized.

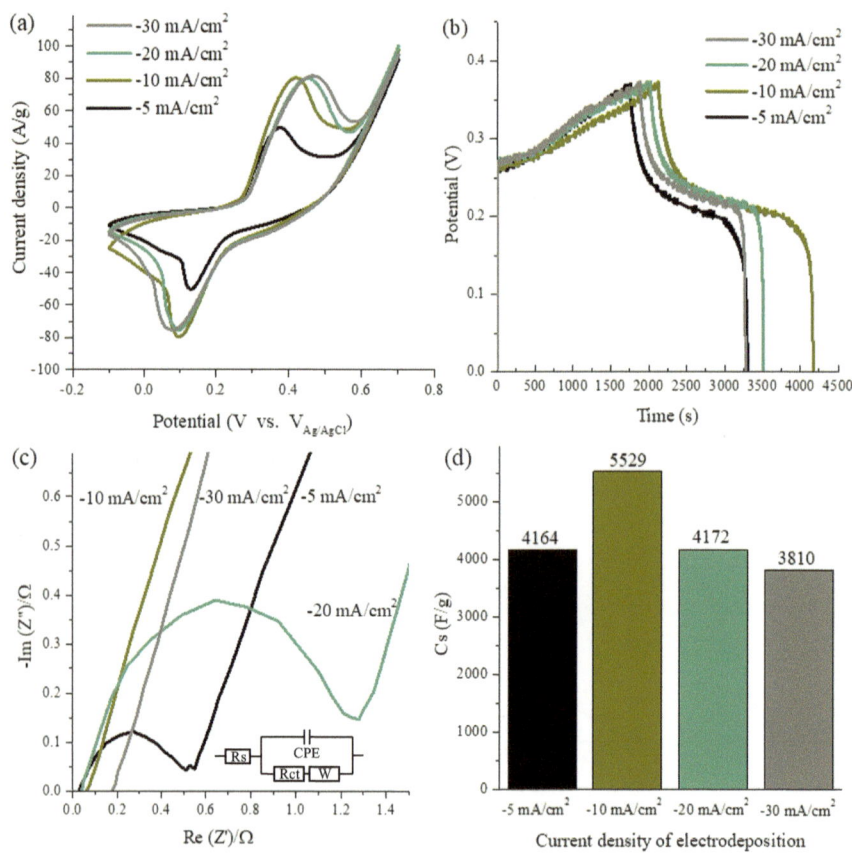

Fig. 3.8 **a** CV, **b** GCD, **c** EIS, and **d** the specific capacities of VN1 cathodes with different current densities at -5, -10, -20, and -30 mA/cm^2 in electrodeposition processes for 5 min. (Elsevier Copyright 2022 with quote number: 501883039)

The unique potential peak seen in Fig. 3.9b, GCD curves align with CV curves, suggesting that VN1 cathodes operate via the Faradaic process. The energy densities of VN1 electrodeposited for 1, 2, 3, 4, 5, and 7 min are 51, 75, 86, 94, 116, and 63 Wh/Kg, respectively, at 1 A/g, based on the discharge profiles. The Nyquist plots of VN1 with various electrodeposition times are shown in Fig. 3.9c. VN1 showed stronger charge transfer resistivities for deposition durations at 3 and 7 min, according to the Nyquist plots. Nonetheless, the Warburg impedances of VN1 cathodes displayed identical slopes in resistivity. Finally, using GCD measurement as a basis, the Cs values of VN1 cathodes with various electrodeposition periods were computed, as shown in Fig. 3.9d. For 1, 2, 3, 4, 5, and 7 min, the obtained specific capacitances were 2438, 3561, 4087, 4453, 5529, and 3007 F/g, respectively. After the best cathode material of electrodeposited VNx has been identified, the optimized VN1 electrode undergoes additional electrochemical tests to examine its characteristics.

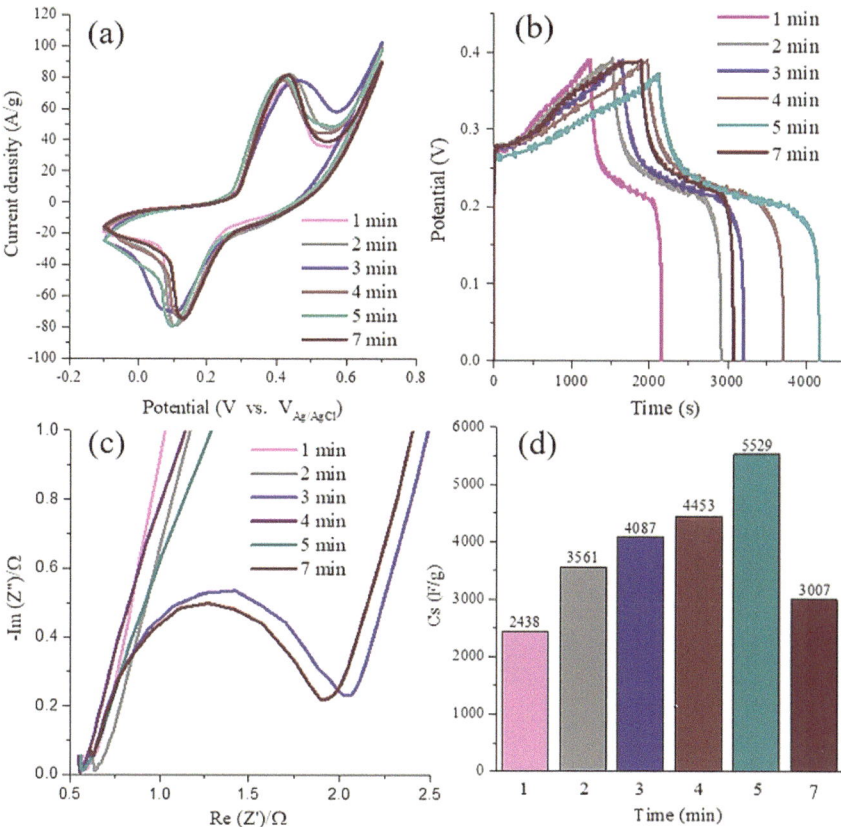

Fig. 3.9 **a** CV, **b** GCD, **c** EIS, and **d** the specific capacities of VN1 cathodes using current density at −10 mA/cm² in the electrodeposition process and variation of deposition times for 1, 2, 3, 4, 5, and 7 min. (Elsevier Copyright 2022 with quote number: 501883039)

The features of the optimized VN1 were investigated using CV and GCD studies, as illustrated in Fig. 3.10. The anodic and cathodic peaks with positive and negative shifts are seen by the CV spectra in Fig. 3.10a, respectively. Figure 3.10b illustrates the results of additional analysis of the peak shift using the square root of the scan rate. The link between scan rate (V/t) and peak current (I) can be expressed as follows: $I = a(v/t)^b$, where b is the equation slope, which varies from 0.5 to 1. The behavior of the electrochemical reaction is mostly an ion insertion diffusion process if $b = 0.5$. It is mostly impacted by the pseudocapacitance behavior during the redox reaction when $b = 1$ (extraction) [12]. GCD measurements of the optimized VN1 developed at various current densities are shown in Fig. 3.10c. The CV curves and the particular potential plateaus support the battery-type storage mechanism of VN1. A second calculation shows that the Cs values of VN1 were 5529, 2902, 2054, 1081, and 486 with applied current densities of 1, 3, 5, 10, and 20 A/g, respectively. In the battery-type electrode, the Cs values are relatively big without Li and Na ions.

A lower *Cs* value is observed as the current density rises because ion diffusion does not have enough time to intercalate deeply to the electroactive material, as seen in Fig. 3.10d [13]. The result indicates that as scan speeds increase, the capacitive current overtakes the diffusion current. With modest scan rates ranging from 0.1 to 0.5 mV/s, the VN1 electrode is subjected to CV testing in order to verify the diffusion and capacitive mechanisms. The diffusive and non-diffusive controlled contributions can be distinguished using Eq. (3.1) [14, 15].

$$I(V) = k_1 V + k_2 V^{1/2} \tag{3.1}$$

Non-diffusion-controlled capacitive current and semi-infinite-diffusion-controlled Faradaic current are denoted by the words $k_1 V$ and $k_2 V^{1/2}$, respectively. Plotting $I/V^{1/2}$ versus $V^{1/2}$ yields the slope value of k_1. Equation (3.1) is used to calculate the results, which indicate that the capacitance contribution outweighs the

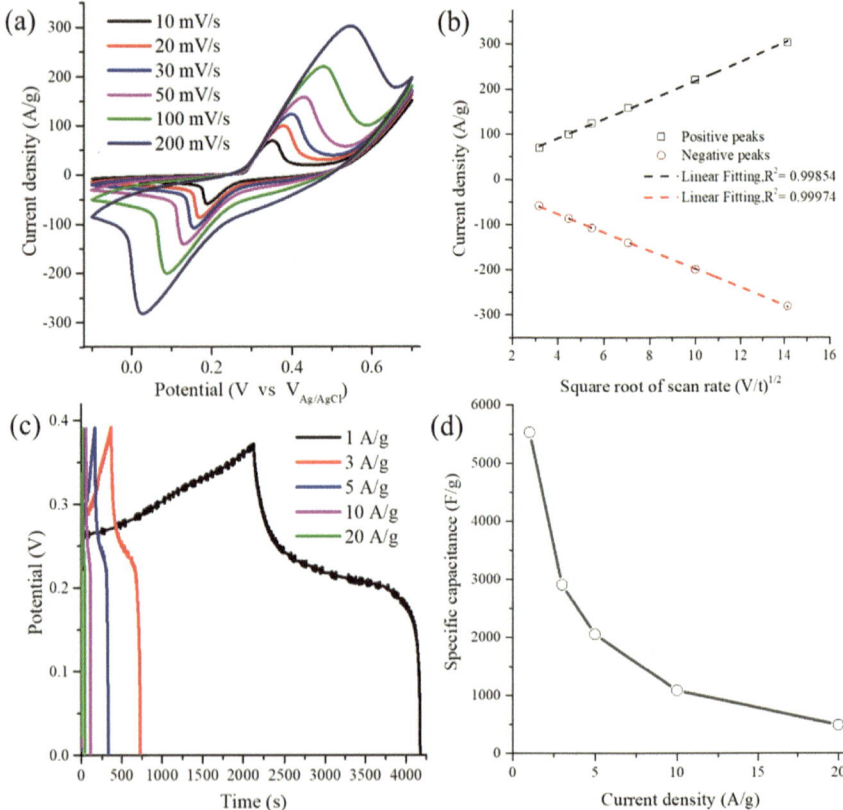

Fig. 3.10 a CV with **b** the corresponding curve of peak currents vs. square root of scan rate, **c** GCD with different current densities, and **d** the plot of different *Cs* values versus current densities. (Elsevier Copyright 2022 with quote number: 501883039)

diffusion contribution at low scan rates of 0.1 to 0.5 mV/s. This is because the capacitive contribution area lies outside of the original CV curve due to the polarization effect [16, 17]. The non-diffusion contribution rises with scan rate, demonstrating the superior capacitance of the electrode. When combined with an excellent energy storage material, the capacitive behavior is preferred and advantageous to the SC performance. The long-term stability of SC material can be attained, the material structure can be preserved without being pulverized or destroyed, and ions typically exist on the surface of electroactive materials because of the non-diffusion-controlled process [18]. However, in order to achieve a reasonable specific capacity of the VN1 electrode, the capacitive or non-diffusion-controlled method is ideal for reducing the ion-transport distance and promoting the charge-transfer rate [19–21].

3.3 Asymmetric Cell of VN1//Active Carbon Supercapacitor Performances

The results of characterizing the asymmetric VN-1/(AC) cell in 3 M KOH electrolyte using the CV, GCD, EIS, and Ragone plot are displayed in Fig. 3.11. The specific capacity of the commercially available AC used in this investigation was determined through chemical characterization. Table 3.5 displays the computed Ed and Pd for various current densities.

Additionally, as illustrated in Fig. 3.12, the stability of the VN-1/AC cell is also examined by utilizing a PVA-KOH gel electrolyte in numerous charge–discharge

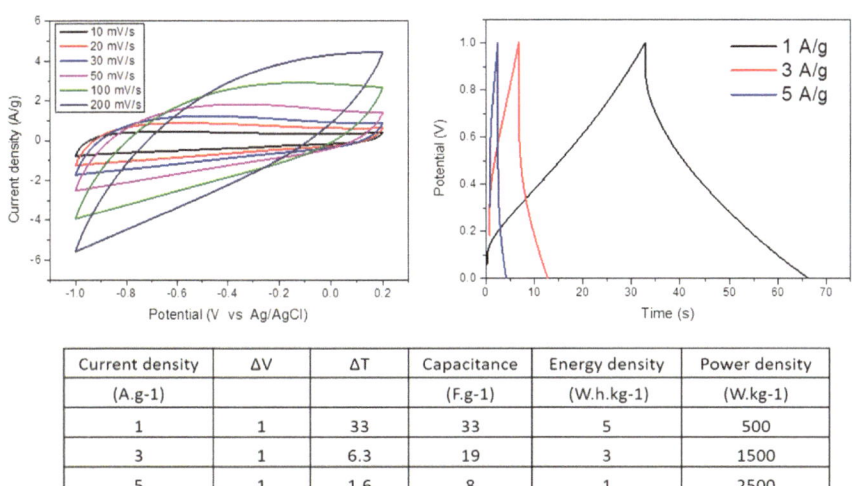

Current density	ΔV	ΔT	Capacitance	Energy density	Power density
(A.g-1)			(F.g-1)	(W.h.kg-1)	(W.kg-1)
1	1	33	33	5	500
3	1	6.3	19	3	1500
5	1	1.6	8	1	2500

Fig. 3.11 Electrochemical properties of commercial active-carbon indicated with CV and GCD analyses and the tabulated data to show its specific capacitances with different current densities. (Elsevier Copyright 2022 with quote number: 501883039)

Table 3.5 Electrochemical performances of VN-1//AC cell in KOH electrolyte

Applied current density (A/g)	V_{max} (V)	E_d (Wh/kg)	P_d (W/kg)
1	1.5	200	1121
3	1.5	95	2250
5	1.5	70	3705
10	1.5	51	7344
20	1.5	33	14,850

processes. In order to keep the SC cell functional and prevent the significant dissolution of vanadium atoms in the KOH solution during charge–discharge cycles, PVA was used to absorb the KOH solution and create a gel electrolyte. Furthermore, there was no deterioration in the gel-type PVA-KOH cell performance after two weeks of storage. The cycling test was conducted up to 10,000 times, confirming the steady phenomenon and maintaining a capacity retention rate of above 100%. As the active-site activation during the cycling test, the retention percentage is not degraded. It is commonly acknowledged that increases in a specific capacity of electrode material are mostly due to factors such as its greater specific surface area, appropriate pore size, and pore size distribution, which give the electrolyte (OH$^-$) efficient channels for mass transport and ion diffusion [22]. Furthermore, the Columbic efficiency stays at 100%, demonstrating the reversible charge–discharge redox reaction.

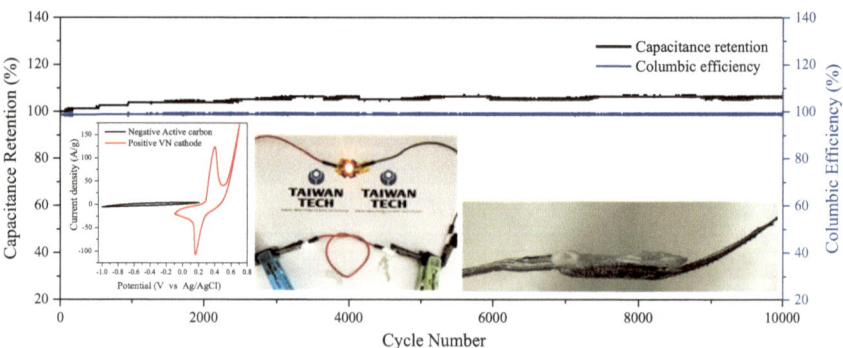

Fig. 3.12 Capacitance retention and Columbic efficiency in the multiple charge–discharge processes for 10,000 cycles with the shown CV of AC and VN electrodes, photographs of serial VN//AC cells for LED lighting, and the VN//AC cell in the inset. (Elsevier Copyright 2022 with quote number: 501883039)

3.4 Stability and Reaction Mechanism of VN1 Cathode in KOH Solution

The VN1 electrode was reexamined using XPS analysis to investigate the elemental chemical states after it had been employed in the cycle test. Following the 10-K cycling test, the binding energy of V in VN1 is displayed in Fig. 3.6d. The structure still contains the oxidation states of V^{4+} and V^{3+}, whose $2p_{3/2}$ and $2p_{1/2}$ orbitals are detected at 514.7, 522.2, 515.7, and 523.3 eV, respectively. The V peak locations in the used VN1 are similar to those of the as-prepared VN-1, as seen in Fig. 3.6a. Nonetheless, a comparatively minor quantity of an extra V metal $2p_{1/2}$ orbital was detected in the spectra at 519.8 eV, suggesting that some V-oxide was converted to V metal during the discharging process [9]. Regarding the Ni XPS spectra, it can be shown that Ni has partially transitioned from its previous chemical state of Ni^{2+} to Ni^{3+}. The new $2p_{3/2}$ and $2p_{1/2}$ orbitals can be seen in Fig. 3.6e at 856.8 and 878.5 eV, respectively [10]. In addition, there is confirmation of the oxygen spectra of O_{ads} and O 1 s orbitals at 532.0 and 528.3 eV, respectively, as presented in Fig. 3.6f. Remarkably, the pseudocapacitive OH-anion adsorption on the VN1 cathode surface improves the adsorbed oxygen substantially following the charge and discharge processes. One possible explanation for the declined V-content in VN1 in the 10 K-cycle tests is that during the charge and discharge operation, V dissolves in a basic PVA/KOH gel solution. 10.13 ppm V were dissolved in the electrolyte after 10 K cycles, according to an inductively coupled plasma (ICP) measurement. As electron extraction during the charging process, the Ni XPS data shows that the Ni^{2+} content decreased to 2.9% and the Ni^{3+} content grew to 5.7%. When the OH^- anion intercalated and covered the surface of VN1 following the cycling test, the number of OH_{ads} also increased to 21.8%.

Some cathode materials with an electrode related to vanadium are included in Table 3.6 for comparison to determine the electrochemical performance of VN1 in this work. Because of the improved active sites, the current study with the VN1 cathode shows a comparatively high specific capacity of 5529 F/g at 1 A/g with stability >100% after 10 K cycles. In comparison to those previous works, the VN1 electrode with electrodeposition method shows a significant improvement as a result.

Vanadium oxide is widely known to dissolve in a basic solution, and charge–discharge procedures may release the V atoms present on the surface of VN1. As a result, the cycling test performance was poor in numerous works utilizing vanadium oxide-based cathodes [37, 38]. The cycling test, however, is extremely stable when the experiment is carried out using a PVA-KOH gel solution, suggesting that even if the surface V atoms dissolved into the solution, V atoms should be reprecipitated on the VN1 surfaces as vanadium oxide. As a result, during the cycling test, the electron transport to the cathode surface can still maintain good stability.

A specific route may be deduced based on the chemical states of Ni as shown in the XPS data in order to better comprehend the mechanism of the highly capacitive VN1 cathode material. Ni comes in two different chemical states: 2^+ and 3^+. This means that when an electron is transferred to the cathode materials, Ni^{2+} will typically

Table 3.6 The electrochemical performance of the VN1 cathode compared to other V-based cathodes

Working electrode	Electrolyte	Window voltage (V)	Performance (F/ g)	Cycling stability (%)	References
Ni-V_3O_5/Ni foam (VN-1)	3 M KOH	0.4	5525 F/g at 1 A/ g	>100% after 10 K cycles	This work
NCV-LDH NSAs	6 M KOH	0.5	2960 F/g at 1 A/ g	81.8% after 10 K cycles	[23]
Fe-VO-S	0.5 M Na_2SO_4	0.8	217 F/g at 3 A/g	92% after 4 K cycles	[24]
V_2O_5	2 M KOH	1.6	205.06 F/g at 2 mV/s	–	[25]
$FeVO_4$	1 M $LiPF_6$	2.0	220 mAh/g at 0.05 A/g	78% after 40 cycles	[26]
Mo–V–ZnO	2 M KOH	1.6	585 F/g at 0.5 A/g	98% after 2 K cycles	[27]
V_2O_5–VO_2/rGO	1 M Na_2SO_4	1.7	468.5 F/g at 1 A/g	89.8% after 10 K cycles	[28]
V_xTe_y/MWCNTs	1 M $LiClO_4$	1.0	470 F/g at 2 mV/s	82.5% after 10 K cycles	[29]
$(NH_4)_2V_{10}O_{25} \cdot 8H_2O$ (NVO)	NH_4Cl/PVA	1.8	324 mF/cm^2 at 1 mA/cm^2	71% after 14 K cycles	[30]
rGO@V_2O_5	5 M $LiNO_3$	0.6	83.5 F/g at 0.5 A/g	73.3% after 1 K cycles	[31]
V_6O_{13}@C	1 M Na_2SO_4	0.7	108.9 F/g at 0.2 A/g	34.5% after 1 K	[32]
VN@C	6 M KOH	1.5	17.5 F/g at 1 A/ g	88.3% after 2 K	[33]
VS_2@Carbon Cloth	1 M Na_2SO_4	0.3	972.5 mF/g at 1 mA/cm^2	95% after 2 K	[34]
$NiCo_2S_4$@NiV-LDH/ NF	6 M KOH	0.5	3557.6 F/g at 1 A/g	77.5% after 10 K	[35]
V-doped Co_3S_4	3 M KOH	1.6	1725 F/g at 0.3 A/g	80% after 5 K	[36]

form to balance the charge in the lattice ($Ni^{3+} + e^- = Ni^{2+}$) during the discharging processes. $Ni^{2+} = Ni^{3+} + e^-$ is the opposite reaction that happens during the charging process when the electron is taken out of the cathode. The intercalation of electrolyte ions (K^+) into the VN1 framework without changing its morphology or structure is what gives the vanadium-based (VN1) cathode its pseudocapacitive nature, as demonstrated by Eq. (3.2) [39].

$$(Ni - V_3O_5)_{surface} + xK^+ + xe^- = (K_xNi - V_3O_5)_{surface} \qquad (3.2)$$

Furthermore, during the discharging process, the K cations can intercalate into the oxyvanite V_3O_5 lattice, providing excellent charge storage. An ICP result confirms that vanadium dissolution in PVA-KOH gel electrolyte makes a small but considerable contribution to the VN1 performance. According to Eqs. (3.3–3.5), equilibrium processes during the charge and discharge conditions may result from the dissolution of V in alkaline circumstances. In alkaline conditions, water and oxygen in the surrounding environment will react to produce hydrogen peroxide, as demonstrated by Eq. (3.2) [40]. The hydrogen peroxide and hydroxyl produced will cause a further electro-Fenton-like reaction, which will dissolve the V-oxide on the electrode [41]. However, the XPS result shows that V metal can be re-deposited and then oxidized on the surface, indicating that the equilibrium processes of V do not have the opposite effect on the long-term capability of VN1 cathode. Because of this, the 10-K cycle test in Fig. 3.12 displays the performance of active charge storage.

$$O_2 + H_2O + 2e^- = HO_2^- + OH^- \tag{3.3}$$

$$V_xO_y + HO_2^- = VO_4^{3-} + OH + OH^- \tag{3.4}$$

$$V_xO_y + OH + e^- = VO_4^{3-} + H_2O \tag{3.5}$$

References

1. D. Chen, H. Tan, X. Rui, Q. Zhang, Y. Feng, H. Geng, C. Li, S. Huang, Y. Yu, Oxyvanite V_3O_5: a new intercalation-type anode for lithium-ion battery. InfoMat **1**(2), 251–259 (2019)
2. W. Demtröder, *Laser Spectroscopy: Vol. 1: Basic Principles*. (Springer Berlin Heidelberg, 2008)
3. M. Shahid, J. Liu, Z. Ali, I. Shakir, M.F. Warsi, Structural and electrochemical properties of single crystalline MoV_2O_8 nanowires for energy storage devices. J. Power. Sources **230**, 277–281 (2013)
4. N.T. Hong Trang, N. Lingappan, I. Shakir, D.J. Kang, Growth of single-crystalline β-$Na_{0.33}V_2O_5$ nanowires on conducting substrate: a binder-free electrode for energy storage devices. J. Power Sources **251**, 237–242 (2014)
5. D. Xiong, W. Li, L. Liu, Vertically aligned porous nickel(II) hydroxide nanosheets supported on carbon paper with long-term oxygen evolution performance. Chem.—Asian J. **12**(5), 543–551 (2017)
6. K. Kishi, K. Fujiwara, Ultrathin nickel oxide on the $V_2O_3Cu(100)$ surface studied by XPS. J. Electron Spectrosc. Relat. Phenom. **71**(1), 51–59 (1995)
7. H. Abdullah, N.S. Gultom, D.-H. Kuo, Depletion-Zone size control of p-type NiO/n-type Zn(O, S) nanodiodes on high-surface-area SiO_2 nanoparticles as a strategy to significantly enhance hydrogen evolution rate. Appl. Catal. B **261**, 118223 (2020)
8. H. Abdullah, D.-H. Kuo, X. Chen, High efficient noble metal free Zn(O, S) nanoparticles for hydrogen evolution. Int. J. Hydrogen Energy **42**(9), 5638–5648 (2017)
9. M.C. Biesinger, L.W.M. Lau, A.R. Gerson, R.S.C. Smart, Resolving surface chemical states in XPS analysis of first row transition metals, oxides and hydroxides: Sc, Ti, V, Cu and Zn. Appl. Surface Sci. **257**(3), 887–898 (2010)

10. J.F. Moulder, J. Chastain, R.C. King, Handbook of X-ray photoelectron spectroscopy: a reference book of standard spectra for identification and interpretation of XPS data. Phys. Electron.: Eden Prairie, MN (1995)
11. C.V.V.M. Gopi, R. Vinodh, S. Sambasivam, I.M. Obaidat, S. Singh, H.-J. Kim, Co_9S_8-Ni_3S_2/$CuMn_2O_4$-$NiMn_2O_4$ and $MnFe_2O_4$-$ZnFe_2O_4$/graphene as binder-free cathode and anode materials for high energy density supercapacitors. Chem. Eng. J. **381**, 122640 (2020)
12. X. Zhang, M. Jin, Y. Zhao, Z. Bai, C. Wu, Z. Zhu, H. Wu, J. Zhou, J. Li, X. Pan, E. Xie, Improved lithium-ion battery performance by introducing oxygen-containing functional groups by plasma treatment. Nanotechnology **32**(27), 275401 (2021)
13. R. Li, S. Wang, Z. Huang, F. Lu, T. He, $NiCo_2S_4$@$Co(OH)_2$ core-shell nanotube arrays in situ grown on Ni foam for high performances asymmetric supercapacitors. J. Power. Sources **312**, 156–164 (2016)
14. F. Zhang, J. Ma, H. Yao, Ultrathin Ni-MOF nanosheet coated $NiCo_2O_4$ nanowire arrays as a high-performance binder-free electrode for flexible hybrid supercapacitors. Ceram. Int. **45**(18, Part A), 24279–24287 (2019)
15. X. Zhang, J. Wang, Y. Sui, F. Wei, J. Qi, Q. Meng, Y. He, D. Zhuang, Hierarchical nickel-cobalt phosphide/phosphate/carbon nanosheets for high-performance supercapacitors. ACS Appl. Nano Mater. **3**(12), 11945–11954 (2020)
16. L. Cheng, Q. Zhang, M. Xu, Q. Zhai, C. Zhang, Two-for-one strategy: three-dimensional porous Fe-doped Co_3O_4 cathode and N-doped carbon anode derived from a single bimetallic metal-organic framework for enhanced hybrid supercapacitor. J. Colloid Interface Sci. **583**, 299–309 (2021)
17. J. Dong, S. Li, Y. Ding, Anchoring nickel-cobalt sulfide nanoparticles on carbon aerogel derived from waste watermelon rind for high-performance asymmetric supercapacitors. J. Alloy. Compd. **845**, 155701 (2020)
18. M.S. Rahmanifar, H. Hesari, A. Noori, M.Y. Masoomi, A. Morsali, M.F. Mousavi, A dual Ni/Co-MOF-reduced graphene oxide nanocomposite as a high performance supercapacitor electrode material. Electrochim. Acta **275**, 76–86 (2018)
19. Y. Xu, Z. Lin, X. Zhong, X. Huang, N.O. Weiss, Y. Huang, X. Duan, Holey graphene frameworks for highly efficient capacitive energy storage. Nat. Commun. **5**(1), 4554 (2014)
20. M. Acerce, D. Voiry, M. Chhowalla, Metallic 1T phase MoS_2 nanosheets as supercapacitor electrode materials. Nat. Nanotechnol. **10**(4), 313–318 (2015)
21. M.-Q. Zhao, X. Xie, C.E. Ren, T. Makaryan, B. Anasori, G. Wang, Y. Gogotsi, Hollow MXene spheres and 3D macroporous MXene frameworks for Na-ion storage. Adv. Mater. **29**(37), 1702410 (2017)
22. D.P. Chatterjee, A.K. Nandi, A review on the recent advances in hybrid supercapacitors. J. Mater. Chem. A **9**(29), 15880–15918 (2021)
23. Z. Wu, D. Khalafallah, C. Teng, X. Wang, Q. Zou, J. Chen, M. Zhi, Z. Hong, Vanadium doped hierarchical porous nickel-cobalt layered double hydroxides nanosheet arrays for high-performance supercapacitor. J. Alloy. Compd. **838**, 155604 (2020)
24. P. Asen, S. Shahrokhian, A.I. Zad, Iron-vanadium oxysulfide nanostructures as novel electrode materials for supercapacitor applications. J. Electroanal. Chem. **818**, 157–167 (2018)
25. D. Velpula, S. Konda, S. Vasukula, S.C. Chidurala, Microwave radiated comparative growths of vanadium pentoxide nanostructures by green and chemical routes for energy storage applications. Mater. Today: Proc. **47**, 1760–1766 (2021)
26. Y. Si, G. Liu, C. Deng, W. Liu, H. Li, L. Tang, Facile synthesis and electrochemical properties of amorphous $FeVO_4$ as cathode materials for lithium secondary batteries. J. Electroanal. Chem. **787**, 19–23 (2017)
27. M.R. Pallavolu, J. Nallapureddy, R.R. Nallapureddy, G. Neelima, A.K. Yedluri, T.K. Mandal, B. Pejjai, S.W. Joo, Self-assembled and highly faceted growth of Mo and V doped ZnO nanoflowers for high-performance supercapacitors. J. Alloy. Compd. **886**, 161234 (2021)
28. Y. Chen, P. Lian, J. Feng, Y. Liu, L. Wang, J. Liu, X. Shi, Tailoring defective vanadium pentoxide/reduced graphene oxide electrodes for all-vanadium-oxide asymmetric supercapacitors. Chem. Eng. J. **429**, 132274 (2022)

29. B. Pandit, S.R. Rondiya, R.W. Cross, N.Y. Dzade, B.R. Sankapal, Vanadium telluride nanoparticles on MWCNTs prepared by successive ionic layer adsorption and reaction for solid-state supercapacitor. Chem. Eng. J. **429**, 132505 (2022)

30. P. Wang, Y. Zhang, H. Jiang, X. Dong, C. Meng, Ammonium vanadium oxide framework with stable NH_4^+ aqueous storage for flexible quasi-solid-state supercapacitor. Chem. Eng. J. **427**, 131548 (2022)

31. H. Liu, W. Zhu, D. Long, J. Zhu, G. Pezzotti, Porous V_2O_5 nanorods/reduced graphene oxide composites for high performance symmetric supercapacitors. Appl. Surf. Sci. **478**, 383–392 (2019)

32. W. Yang, J. Zeng, Z. Xue, T. Ma, J. Chen, N. Li, H. Zou, S. Chen, Synthesis of vanadium oxide nanorods coated with carbon nanoshell for a high-performance supercapacitor. Ionics **26**(2), 961–970 (2020)

33. M.R. Pallavolu, Y. Anil Kumar, R.R. Nallapureddy, H.R. Goli, A. Narayan Banerjee, S.W. Joo, In-situ design of porous vanadium nitride@carbon nanobelts: a promising material for high-performance asymmetric supercapacitors. Appl. Surf. Sci. **575**, 151734 (2022)

34. M.Y. Zhang, J.Y. Miao, X.H. Yan, Y.H. Zhu, Y.L. Li, W.J. Zhang, W. Zhu, J.M. Pan, M.S. Javed, S. Hussain, Vanadium disulfide nanosheets loaded on carbon cloth as electrode for flexible quasi-solid-state asymmetric supercapacitors: energy storage mechanism and electrochemical performance. J. Mater. Chem. C **10**(2), 640–648 (2022)

35. J. Pan, S. Li, F. Li, W. Zhang, D. Guo, L. Zhang, D. Zhang, H. Pan, Y. Zhang, Y. Ruan, Design and construction of core-shell heterostructure of Ni–V layered double hydroxide composite electrode materials for high-performance hybrid supercapacitor and L-Tryptophan sensor. J. Alloy. Compd. **890**, 161781 (2022)

36. E. Niknam, H. Naffakh-Moosavy, S.E. Moosavifard, M. Ghahraman Afshar, Amorphous V-doped Co_3S_4 yolk-shell hollow spheres derived from metal-organic framework for high-performance asymmetric supercapacitors. J. Alloy. Compd. **895**, 162720 (2022)

37. I. Shakir, Z. Ali, J. Bae, J. Park, D.J. Kang, Layer by layer assembly of ultrathin V_2O_5 anchored MWCNTs and graphene on textile fabrics for fabrication of high energy density flexible supercapacitor electrodes. Nanoscale **6**(8), 4125–4130 (2014)

38. D. Majumdar, M. Mandal, S.K. Bhattacharya, V_2O_5 and its carbon-based nanocomposites for supercapacitor applications. ChemElectroChem **6**(6), 1623–1648 (2019)

39. J.S. Daubert, N.P. Lewis, H.N. Gotsch, J.Z. Mundy, D.N. Monroe, E.C. Dickey, M.D. Losego, G.N. Parsons, Effect of meso- and micro-porosity in carbon electrodes on atomic layer deposition of pseudocapacitive V_2O_5 for high performance supercapacitors. Chem. Mater. **27**(19), 6524–6534 (2015)

40. Y. Xue, Y. Wang, S. Zheng, Z. Sun, Y. Zhang, W. Jin, Efficient oxidative dissolution of V_2O_3 by the in situ electro-generated reactive oxygen species on N-doped carbon felt electrodes. Electrochim. Acta **226**, 140–147 (2017)

41. S. Chen, J. Hong, H. Yang, J. Yang, Adsorption of uranium (VI) from aqueous solution using a novel graphene oxide-activated carbon felt composite. J. Environ. Radioact. **126**, 253–258 (2013)

Chapter 4
Electrodeposited VO_x on $Ni(OH)_2$ Layer Grown on Ni Foam Substrate for High-Charge Storage Supercapacitor (Adapted with Permission from {23}. Copyright {2023} American Chemical Society)

4.1 Ni- and V-Based Electrode Materials as the Promising Candidates for Electrode Materials

Some important works on Ni-based materials have been done to improve SCs [1–4]. The present progress in SC research has been improved by noteworthy findings. The asymmetric flexible Ni–Co sulfide nanosheet/CNT hybrid film with a power density of 900/18000 W/Kg with an energy density of 59.5/34.5 Wh/Kg and an output voltage of 1.8 V was created by Lu et al. In addition, after 10,000 cycles, the capacity retention can increase to 86.64% [5]. A flexible SC of $Ni(OH)_2$@Ni core–shell electrode with good cycle stability and a *Cs* of 2454 F/g at 5 A/g current density was produced by Su et al. [6] $NiFeP@NiCo_2S_4$ nanosheet array was produced by Wan et al. on carbon cloth for asymmetric SC, and after 5000 cycles, the capacitance was 874.4 F/ g at 1 A/g with a capacity retention rate of 85.6% [7]. Furthermore, a stable sulfide-based composite with a NiS@CoS core–shell that demonstrated a high capacitance of 1210 F/g at 1 A/g was also employed by Miao et al. At 10 A/g, the NiS@CoS core–shell electrode's capacity retention might reach 82% [8]. Guan et al. also discovered the enhanced SC performances achieved by employing NiS on MoS_2 nanosheet arrays. The NiS outstanding flexibility and charge–discharge stability might greatly increase the specific capacity [9]. Previous studies have consistently shown that Ni-based electrodes are necessary components for SCs with comparatively high specific capacitance.

In addition to Ni-based electrodes, V-based materials have also been thoroughly investigated for energy-storage technologies due to their abundance, different valences, low cost, and typical layered structure [10]. Combining these two kinds of materials is interesting and promising for applications. The general limitations of VO_x materials are inferior electrochemical performances and charge–discharge process instability [11]. Nevertheless, several tactics were implemented to surmount the limitations of VO_x materials. In their study, Wang et al. synthesized a nanoflake composite of V_2O_5 and Ni_3S_2, achieving a capacity retention of 85% after 2500 cycles

© The Author(s), under exclusive license to Springer Nature Singapore Pte Ltd. 2024   57
H. Abdullah, *Vanadium Oxide-Based Cathode for Supercapacitor Applications*,
SpringerBriefs in Applied Sciences and Technology,
https://doi.org/10.1007/978-981-97-5243-0_4

and a *Cs* value of 3060 F/g [12]. N-doped graphene-composited VN has been done by Balamurugan et al. for the SC device. After 10,000 cycles at 10 A/g, it attained a specific capacitance of 445 F/g with a retention of 99% [13]. By combining V$_2$O$_5$ and N-doped graphene aerogel, Wei et al. were able to create a SC device with a capacity retention of 95% after 20,000 cycles and a *Cs* of 710 F/g at 0.5 A/g [14]. In another alternative methodology, Shakir et al. fabricated a 2590 F/g capacitance using a V$_2$O$_5$ thin layer coated on MWCNT and graphene, which retained 96% of its original value after 5000 cycles [15]. Some characteristics of V-based electrode materials are useful for supercapacitor devices that resemble batteries and have been utilized to increase battery performance [16]. By leveraging the battery and SC features of V-based material, which is present in some energy storage devices, it is possible to advance energy storage technology in a diagonal direction toward greater power and energy densities in the Ragone plot.

   In this chapter, an effort to increase the Cs value was made by growing an amorphous Ni-doped VO$_x$ on Ni(OH)$_2$ lamellar structure on Ni foam (NF). The simple cathode preparation consists of two straightforward procedures: (*I*) aging the NF for four hours in nitric acid at 80 °C and (II) electrodepositioning a-Ni-doped VO$_x$. The application of a Ni(OH)$_2$ lamellar structure to the amorphous VO$_x$ phase was verified. Elevated Faradaic charge storage was seen in the half-cell performance of the as-prepared cathode [17]. The computed power and energy densities have the potential to attain 199 W/kg and 167 Wh/Kg, correspondingly.

4.2   Fabrication and Characterization Methods of Amorphous Ni-Doped VO$_x$/Ni(OH)$_2$ Lamellar on Ni Foam Substrate

Fabrication of amorphous Ni-doped VO$_x$/Ni(OH)$_2$ lamellar (denoted as NVO) involves a two-step method: growing the Ni(OH)$_2$ lamellar on Ni foam (NF) and electrodepositing VO$_x$ with Ni dopant on the Ni(OH)$_2$ layer. Ni foam as the substrate is cut into the size of 1 cm × 2 cm and treated by immersing it in acetone and ethanol for 30 min and drying in an oven for 2 h. As illustrated in Fig. 4.1, the as-cleaned NF was submerged in 0.1 M HNO$_3$ at 80 °C for four hours to develop the lamellar structure. During the heating process, the Ni surface is oxidized to form Ni(OH)$_2$ lamellar. In the electrodeposition process, the as-grown Ni(OH)$_2$ lamellar was utilized as a working electrode with 50 mL electrolyte consisting of 0.05 M VCl$_3$ and 0.05 M NiCl$_2$. Electrodeposition was performed at a Bio-Logic Science workstation with Ag/AgCl and Pt plates serving as the reference and counter electrodes, respectively. For a variety of NVO layer deposition times on the substrate, the NF substrate was subjected to a continuous current of -30 mA as the working electrode for 1, 3, 5, and 7 min. After the electrodeposition process, the electrodes were dried in an oven for 6 h. The mass of deposited chemicals was obtained from the scaling before and after the

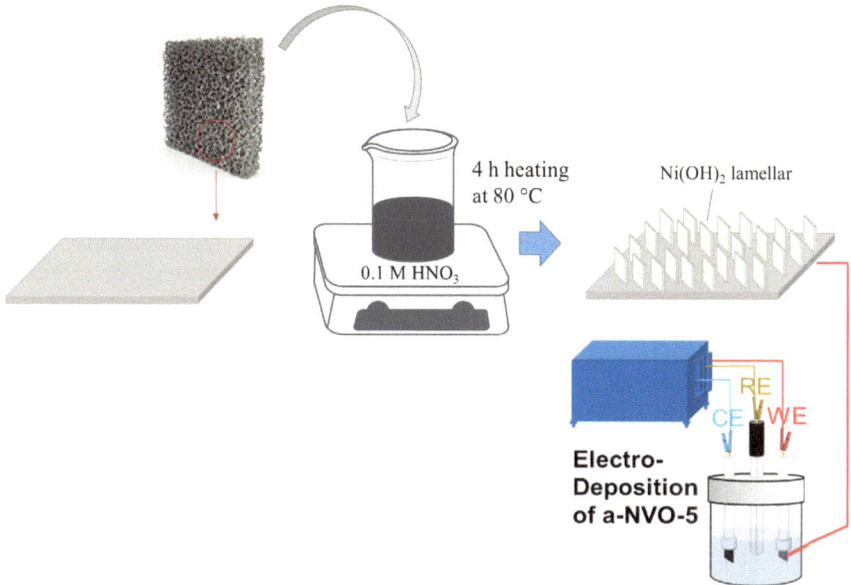

Fig. 4.1 Experimental steps in growing Ni(OH)$_2$ lamellar on Ni foam substrate

drying process. For electrodeposition performed at 1, 3, 5, and 7 min, the as-prepared cathodes are designated NVO-1, NVO-3, NVO-5, and NVO-7, respectively.

A series of electrochemical (EC) analyses was done on the as-synthesized NVO-x ($x = 0, 1, 3, 5$, and 7) cathodes. CV, EIS, and GCD measurements are conducted in a 3 M KOH solution. The NVO-x electrodes are employed as the working electrodes throughout the measurement process, whereas the counter and reference electrodes are Pt foil and Ag/AgCl/KCl, respectively. The EC analysis was conducted utilizing a Bio-Logic Science workstation equipped with EC-Lab software. Specific capacitances, energy density, and power density of NVO-x electrodes were calculated based on Cs, Ed, and Pd using the Eqs. (4.1, 4.2, and 4.3), in which m, ΔV, and Δt represented the loading mass of active materials on NF substrates, the potentials during GCD measurement, and discharging time, respectively.

$$Cs = \frac{I}{m\frac{\Delta V}{\Delta t}}\left[\text{F/g}\right] \tag{4.1}$$

$$E_d = \frac{Cs(\Delta V)^2}{7.2}\left[\text{Wh/Kg}\right] \tag{4.2}$$

$$P_d = (E_d/\Delta t) \times 3600\left[\text{W/Kg}\right] \tag{4.3}$$

In addition, the full-cell supercapacitor is built in NVO-x//active carbon (AC) configuration assembled using KOH-PVA gel as the electrolyte. The preparation of active carbon was conducted based on the previous work [18]. To fabricate the anode of full-cell SC, a standard synthesis procedure was applied by combining 0.125 g of PVDF, 80 g of AC (1 g), and 10 g of carbon black in a 10% NMP solution. A Ni foam with a size of 1 cm × 2 cm was submerged partially in the NMP suspension to create a coating area of 1 cm × 1 cm; the NF was then allowed to dry overnight. In order to make the KOH-PVA gel as an electrolyte, 4 g PVA was dissolved in 50 mL DI water at 90 °C. After the PVA solution was cooled down to room temperature, 4 mL of 12 M KOH solution was subsequently added to it. As the charge storage capacity is different for each material. The mass balance of the cathode and anode should be adjusted and predicted based on Eq. (4.4).

$$\frac{m_+}{m_-} = \frac{\Delta V_- \Delta C_{S-}}{\Delta V_+ \Delta C_{S+}} \tag{4.4}$$

Achieving maximum charge storage equality between the cathode and anode is the objective of mass balance. Equation (4.4) represents m (g) as the mass of active material, ΔV (V) as the potential window, and C_S (F/g) as the specific capacitance.

A Bruker D2-phaser diffractometer was employed to examine the crystal structures of the as-deposited electrodes at a 1.54060 Ω Cu–Kα wavelength. The elemental mapping and morphology analysis on the NVO-x cathodes were examined with an FE-SEM (field-emission scanning electron microscope, JSM 6500F, JEOL, Tokyo, Japan). Further analysis of phases on NVO-x cathodes was conducted with high-resolution images and lattice fringes, which were probed in a FEG-TEM (field-emission gun transmission electron microscope, Philips Tecnai F30, USA). Raman spectroscopy analysis was applied to prove the existence of VO$_x$. The chemical states and composition of elements in NVO$_x$ cathodes were probed with X-ray photoelectron spectroscopy (XPS) measurements using a Thermo VG scientific spectroscope (ESCALAB 250, England). The Electrochemical analyses including electrochemical impedance spectroscopy (EIS), galvanostatic charge–discharge (GCD), cyclic voltammetry (CV), and stability test for capacity retention and Coulombic efficiency were examined with a reliable SP-300 Bio-Logic science instrument.

4.3 Identification of NVO Cathodes with X-Ray Diffraction Pattern and Raman Spectroscopy

The XRD patterns depicted in Fig. 4.2a enable the identification of the phases of NVO-x on NF substrates. Although NVO cathodes are doped with different amounts of Ni, the XRD patterns of NVO-x (x = 1, 3, 5, and 7) have a similar pattern. For a comparison purpose, the as-grown Ni(OH)$_2$ lamellar on Ni-foam substrate is examined in the XRD analysis. The peaks at 44° and 51° in a-NVO-x/Ni(OH)$_2$, which are confirmed with PDF #03–1051 are attributed to the Ni metal from Ni foam substrate. The peaks with the 2θ located at 19.5°, 33.5°, 38.8°, and 52.2° are matched with the β-Ni(OH)$_2$ phase in NVO-x/Ni(OH)$_2$ on Ni foam. Surprisingly, the XRD analysis does not show the peaks of VO_x phase on Ni foam substrate. Furthermore, Raman analysis is conducted and the spectra are indicated in Fig. 4.2b. The Raman spectra were analyzed at wavenumbers between 130 and 1040 cm^{-1} and 3552 and 3600 cm^{-1}, since no discernible peak was observed in the remaining wavenumber range. The NF substrate displays no discernible peaks during the Raman investigation. In contrast, the broad peak at 3583 cm^{-1} observed when $Ni(OH)_2$ is grown on NF can be attributed to the symmetric stretching of the O–H group in $Ni(OH)_2$. The peak observed at 440 cm^{-1} corresponds to the A_{1g} (T) mode of stretching vibration in Ni–OH [19]. The Eg (T) mode of $Ni(OH)_2$ is further demonstrated by the peak at 304 cm^{-1}, which corresponds to the XRD analysis and signifies the existence of β-$Ni(OH)_2$ [20, 21]. A broad peak at 855 cm^{-1} appears when VO_x is placed on an NF substrate; this peak indicates the amorphous phase of VO_x. The peak corresponds to the bending vibration of V–O–V [22]. On the NF substrate, all associated peaks of $Ni(OH)_2$ and amorphous VO_x emerged concurrently for a-VO_x/$Ni(OH)_2$. The Raman analysis provides confirmation that the cathode in its as-prepared state comprises a-VO_x and $Ni(OH)_2$, arranged in the configuration of a-VO_x/$Ni(OH)_2$ on the NF substrate.

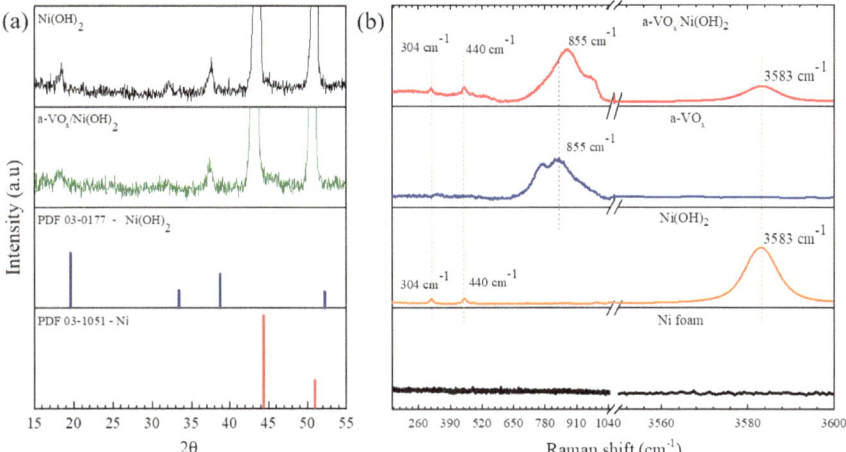

Fig. 4.2 a XRD patterns of $Ni(OH)_2$ and VO_x/$Ni(OH)_2$ compared to the standard PDF files of $Ni(OH)_2$ (PDF #03–0177) and Ni metal (PDF #03–1051) as well as **b** the Raman spectra of Ni foam, $Ni(OH)_2$/Ni foam, VO_x/Ni foam, and VO_x/$Ni(OH)_2$/Ni foam. (Reprinted with permission from [23] Copyright 2023 American chemical society)

4.4 NVO-x Cathode Surface Morphology Examination

VO$_x$/Ni(OH)$_2$ is examined with scanning electron microscopy to determine the surface morphology. The morphologies of Ni(OH)$_2$ lamellar, NVO-1, NVO-3, NVO-5, and NVO-7 with NF as the substrate are indicated in Fig. 4.3. As shown in Fig. 4.3a, the Ni(OH)$_2$ flakes are produced perpendicularly on NF substrate. As illustrated in Fig. 4.3b–e, after the electrodeposition operations at various time intervals, the Ni(OH)$_2$ is coated by secondary layers of VO$_x$. The analysis shows that increasing the amounts of VO$_x$ second layer totally covered the surfaces of Ni(OH)$_2$ in NVO-7. Furthermore, it is evident that NVO-x (where $x = 1$, 3, and 5) possesses porous surfaces that are highly conducive to the storage of charges. As indicated in Fig. 4.3f, the EDS analysis suggests all the elements in NVO-5 are present. The surface composition of NVO-5 was determined to be 17.2% V, 16.5% Ni, and 66% O, as determined by semiquantitative EDS.

Additional examination of the morphology of NVO-5 was conducted utilizing transmission electron microscopy (TEM). A concentrated ion beam was utilized to perform a cross-sectional cut on the NVO-5 cathode prior to the TEM session (FIB). The cross-section of a-NVO-5 on NF substrate is illustrated in Fig. 4.4a. The dark region in Fig. 4.4a corresponds to the Ni foam substrate, whereas the lamellar configuration represents Ni(OH)$_2$. In order to visually depict the interfaces between amorphous-VO$_x$ and Ni(OH)$_2$, a high-resolution transmission electron microscopy (TEM) picture is utilized to analyze a minuscule region as illustrated in Fig. 4.4b. The interfaces of the Ni(OH)$_2$ and a-VO$_x$ phases are visible in the picture. Approximately, 2.32 Å lattice fringes were identified as the lattice parameter of d(101) in

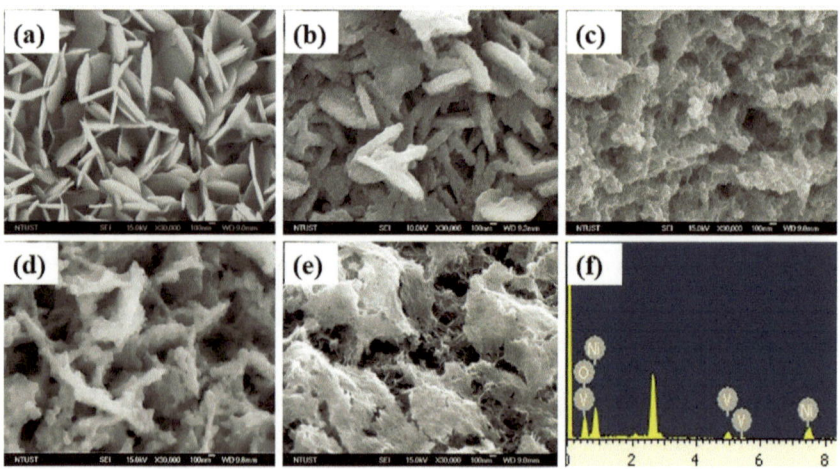

Fig. 4.3 Surface morphology of **a** Ni(OH)$_2$ lamellar, **b** NVO-1, **c** NVO-3, **d** NVO-5, and **e** NVO-7 on NF and **f** the EDS spectrum of NVO-5. (Reprinted with permission from [23] Copyright 2023 American chemical society)

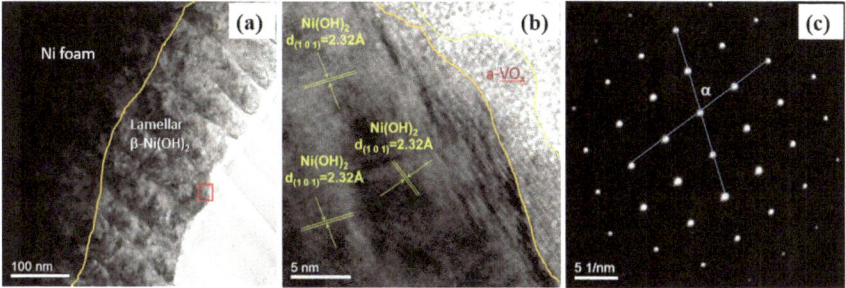

Fig. 4.4 Cross-sectional image of **a** NVO-5, **b** high-resolved Ni(OH)$_2$/VO$_x$ interfaces, and **c** the SAED dot pattern of HCP Ni(OH)$_2$ structure. (Reprinted with permission from [23] Copyright 2023 American chemical society)

the Ni(OH)$_2$ phase. The subsequent amorphous layer of VO$_x$ is likewise visible; yet, the lattice fringes are imperceptible. Furthermore, the pattern generated by selected area electron diffraction (SAED) is readily apparent in Fig. 4.4c. The dot pattern represents the hexagonal close pack (HCP) structure of ς-Ni(OH)$_2$, where the angle α is 60° [24]. The SAED pattern has not been identified for VO$_x$ due to its amorphous nature.

4.5 Surface Chemical State and Composition Analysis

By utilizing XPS measurement, the chemical states and composition of the NVO-5 cathode were determined. The high-resolution XPS spectra of the Ni 2p, O 1 s, and V 2p orbitals in NVO-5 are depicted in Fig. 4.5. The orbitals of Ni^{2+} 2p$_{3/2}$ and 2p$_{1/2}$ were observed at 855.65 and 873.18 eV, respectively, as seen in Fig. 4.5a [25, 26]. A lesser quantity of Ni metal was also produced as a result of the reductive electrodeposition. A lower-intensity binding energy of 850.31 eV was determined. The Ni 2p satellite peaks are detected at 862.30 and 879.63 eV. The binding energies at O 1 s are illustrated in Fig. 4.5b as 531.5 and 532.9 eV, [27, 28] related to the lattice oxygen (O$_L$) and adsorbed OH (OH$_{ads}$) on NVO-5 surfaces, respectively. NVO-5 is indicated in states of V^{3+} and V^{4+}. The binding energies of the orbitals 2p$_{3/2}$ and 2p$_{1/2}$ for V^{3+} in Fig. 4.5c are measured to be 514 and 522.6 eV, correspondingly [29]. Furthermore, the binding energies of 515.4 eV and 523.6 eV for the 2p$_{3/2}$ and 2p$_{1/2}$ orbitals, respectively, validate the chemical state of V^{4+} [30]. The XPS elemental composition is ascertained and presented in Table 4.1, with measurements of the peak area for each orbital in NVO-5. It is determined that 15.5% and 1.5%, respectively, of Ni^{2+} are present in Ni(OH)$_2$ and Ni^0. An oxyvanite composition with an amorphous structure could be suggested by the nearly 2:1 ratio of V^{3+} to V^{4+}. VO$_x$ and Ni(OH)$_2$ contain an estimated 49.91% of the lattice oxygen (O$_L$). The as-prepared cathode surfaces exhibit an estimated 16.19% adsorption of oxygen as a hydroxyl group.

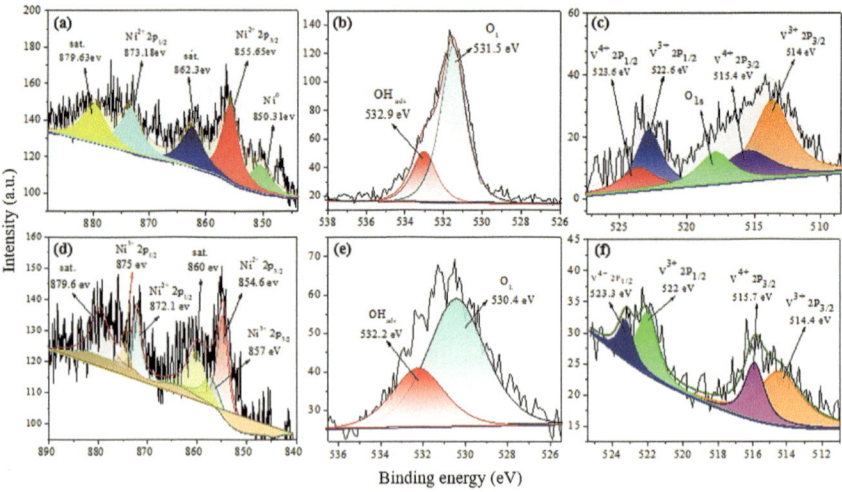

Fig. 4.5 High-resolution XPS spectra of Ni 2p, O 1 s, and V 2p orbitals in (**a, b, c**) freshly prepared and (**d, e, f**) after 10 K-cycle NVO-5 cathodes. (Reprinted with permission from [23] Copyright 2023 American chemical society)

Table 4.1 XPS composition analysis of freshly prepared and after 10 K-cycle NVO-5 cathode

Electrode sample	V (at.%)		Ni (at.%)			O (at.%)	
	V^{4+} (%)	V^{3+} (%)	Ni^{2+} (%)	Ni^{3+} (%)	Ni^0 (%)	O$_{ads}$ (%)	O$_L$ (%)
As-prepared a-NVO-5	5.6	11.3	15.5	–	1.5	16.19	49.91
10 K-cycle a-NVO-5	4	8	12.7	5.3	–	34.5	35.5

4.6 Electrochemical Properties of NVO Cathodes

CV, GCD, and EIS analyses are utilized to estimate the specific capacitance (*Cs*), energy density (*Ed*), and power density (*Pd*) of the electrochemical characteristics of as-prepared NVO-*x* cathodes (*x* = 1, 3, 5, and 7). The Faradaic charge transfer behaviors of NVO-*x* (*x* = 1, 3, 5, and 7) are illustrated in Fig. 4.6a via a cycle of oxidation and reduction peaks [17]. The CV is done with a measurement scan rate of 30 mV/s from −0.9–1.3 V. According to the CV plots, NVO-5 demonstrates the greatest integration area and the most prominent current density peaks in comparison to other NVO-*x* cathodes. The specific capacitance of a charge storage substance is proportional to its integral area [31]. The increased number of electroactive sites at the electrode/electrolyte interface is responsible for the higher redox peaks observed in NVO-5. The galvanostatic charge and discharge (GCD) curves of NVO-x cathodes with a current density of 1 A/g are depicted in Fig. 4.6b. These cathodes are consistent with CV results and have a battery-like appearance. The maximal potential of the GCD curve achieves 0.4 V with a similar charging and discharging time.

The NVO-5 material exhibits the longest charging and discharging times on the GCD curve, suggesting that the charge stored on its active cathode surfaces is more reversible. Equation (4.1) can be utilized to determine the particular capacitances of NVO-x in accordance with the GCD results; the resulting values are illustrated in Fig. 4.6c. The Cs values of NVO-x ($x = 1$, 3, 5, and 7) are 3624, 5377, 7500, and 2710 F/g, respectively. The amorphous NVO-5 cathode demonstrates the greatest Cs value, approaching 167 Wh/Kg for energy density and 199 W/Kg for power density. Additional EIS analysis, as shown in Fig. 4.6d, reveals that at the interface, all NVO-x cathodes have low charge transfer resistances and strong dielectric polarization characteristics. A greater gradient in the low-frequency range signifies a diminished Warburg impedance, which in turn suggests enhanced kinetics of ionic transport and diffusion.

Furthermore, the CV and GCD measurements of NVO-5 were conducted at current densities ranging from 1 to 20 A/g and scan speeds ranging from 10 to 100 mV/s, respectively. As the scan rate increased, the peaks of current densities increased

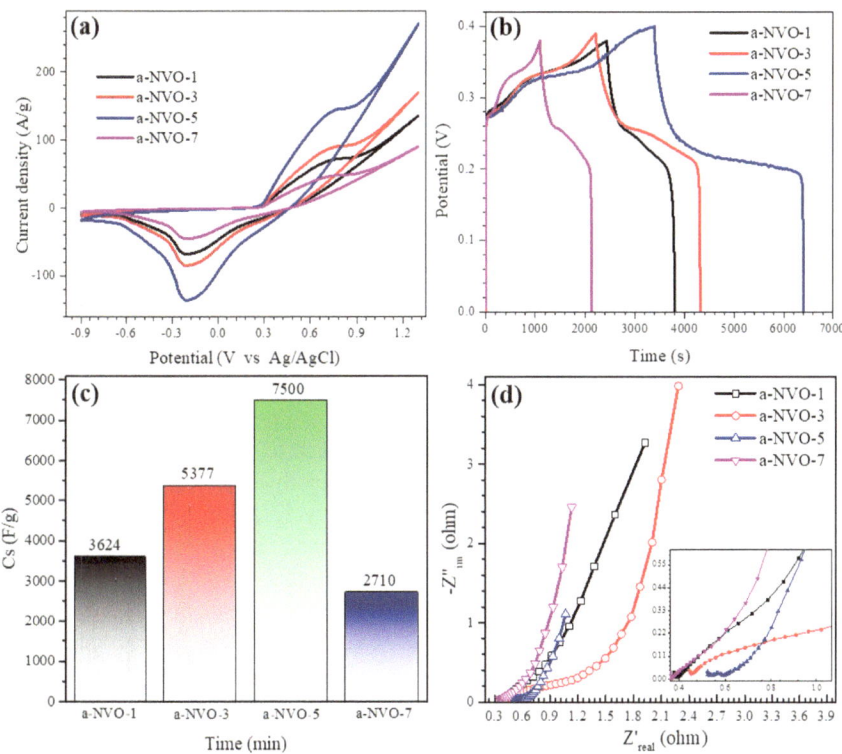

Fig. 4.6 **a** CV, **b** GCD curves, **c** specific capacitances (Cs), and **d** EIS spectra of NVO-x ($x = 1$, 3, 5, and 7) in 3 M KOH solution (Reprinted with permission from [23] Copyright 2023 American chemical society)

linearly, as seen by the CV in Fig. 4.7a. The observed charge storage exhibits pseudocapacitive characteristics at relatively high applied scan rates, indicating that the electrochemical reaction is governed by a reversible Faradaic hydroxyl-ion diffusion process. The NVO-5 GCD curve for various applied current densities ranging from 1 to 20 A/g is illustrated in Fig. 4.7b. As the current density increases, the charge and discharge periods become shorter. Furthermore, a battery-like storing mechanism may account for the increased energy density at a decreased current density, as suggested by the specific potential plateaus. In comparison to the other NVO-x electrodes, the NVO-5 shows a comparatively extended discharging duration. The Cs values for a-NVO-5 with $I/m = 1, 3, 5, 10$, and 20 A/g are 7500, 3825, 3237, 2325, and 550 F/g, respectively, as determined from the GCD curves. At 20 A/g, the Cs value is comparatively high, suggesting that NVO-5 is a suitable energy storage material. As illustrated in Fig. 4.7c, in order to assess the cyclability of NVO-5, the capacitance retention and Coulombic efficiency were monitored for a total of 10 K cycles in a 3 M KOH solution. The capacitance retention increases by almost 100% after 4000 cycles, suggesting that the storage sites of NVO-5 are activated. It is widely acknowledged that the enhanced specific capacitance of SC materials can be ascribed to their larger specific surface area and pore size distribution. These characteristics enable the electrolyte (OH$^-$) to traverse efficiently, hence facilitating ion diffusion and mass transfer [32]. Capacitance retention ultimately reached 93.1% following 10,000 cycles of charge–discharge operations. Furthermore, an unprecedented Coulombic efficiency of 100% was achieved throughout the cyclability procedure.

Following a 10-K cycle test, NVO-5 was subjected to additional XPS analysis. The high-resolution XPS spectra were presented in Fig. 4.5d, e and f, while Table 4.1 detailed the composition analysis. After the cyclability test, the oxidation number of Ni atoms in NVO-5 changed to the Ni^{2+} and Ni^{3+} states. The state of Ni^{2+} is characterized by the binding energies of its 2p$_{3/2}$ and 2p$_{1/2}$ orbitals, which are 854.6 and 872.1 eV, respectively. Additionally, at 857 and 875 eV, distinct Ni^{3+} states are detected; these states correspond to the Ni 2p$_{3/2}$ and 2p$_{1/2}$ orbitals, respectively. In freshly synthesized NVO-5, the Ni metal is observed to be oxidized, as there is no discernible peak at 850.31 eV. Moreover, the XPS analysis indicates that the O$_L$ and O$_{ads}$ remain accessible. During charge–discharge cycles, however, the area ratio of O$_{ads}$ is substantially increased as more hydroxyl ions are adsorbed onto the NVO-5 electrode. The observation of V^{3+} and V^{4+} indicates that the states of V in a-NVO-5 remain unchanged. The state of V^{3+} is indicated by the binding energies of the 2p$_{3/2}$ and 2p$_{1/2}$ orbitals, which are 514.4 and 522.0 eV, respectively. Furthermore, at 515.7 and 523.3 eV, the state of V^{4+} is detected, which corresponds to the 2p$_{3/2}$ and 2p$_{1/2}$ orbitals of V, respectively. The composition analysis presented in Table 4.1 reveals that the 10-K cyclability test resulted in a 2.8% decrease in Ni^{2+} concentrations, a 5.3% increase in Ni^{3+} concentrations, and complete oxidation of Ni^0. The observation with Ni oxidation numbers change during charge–discharge cycles suggests the presence of pseudocapacitance characteristics. The adsorption of OH-ions throughout the electrochemical process is confirmed by the increased concentration of O$_{ads}$ to 34.5%, which is over 100% greater than the concentration before the cyclability test.

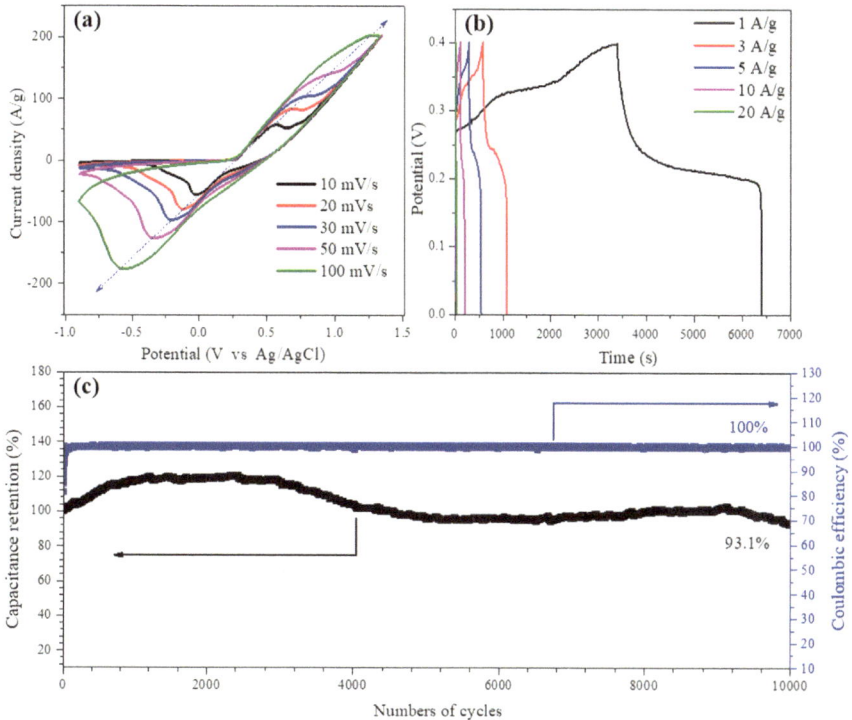

Fig. 4.7 **a** CV, **b** GCD, **c** capacitance retention, and Coulombic efficiency of NVO-5 during stability test in 3 M KOH (Reprinted with permission from [23] Copyright 2023 American chemical society)

The drop in V^{4+} and V^{3+} concentrations following the cyclability test indicates that V species have dissolved in the KOH solution. The capacitance retention reached 93% at 20 A/g despite V dissolution occurring during the test, as shown in Fig. 4.7c, indicating that the dissolved V species may be reprecipitated on NVO-5 and involved in charge–discharge processes.

In order to verify both diffusion and non-diffusion mechanisms, a reduced scanrate CV measurement ranging from 0.1 to 0.5 mV/s is applied to the NVO-5 cathode. Equation (3.1) clarifies the controlled contributions, both diffusive and non-diffusive [33, 34].

The $k_1 V$ and $k_2 V^{1/2}$ variables are the non-diffusion-controlled (capacitive) current and diffusion-controlled faradaic current, respectively. The slope value of k_1 can be obtained by plotting $I/V^{1/2}$ versus $V^{1/2}$. Equation (3.1) indicates that when the scan rate increases, the capacitive contribution becomes more pronounced. The percentages of capacitive and diffusion behaviors during the CV measurement from 0.1 to 0.5 mV/s are displayed in Fig. 4.8a. As the scan rate rose from 0.1 to 0.5 mV/s, the capacitive behavior increased by 27.3% to 65%, as determined by the calculation using CV spectra. Capacitive behavior is favored in the material design for SC because of its advantageous nature in supporting the reversible charge storage mechanism.

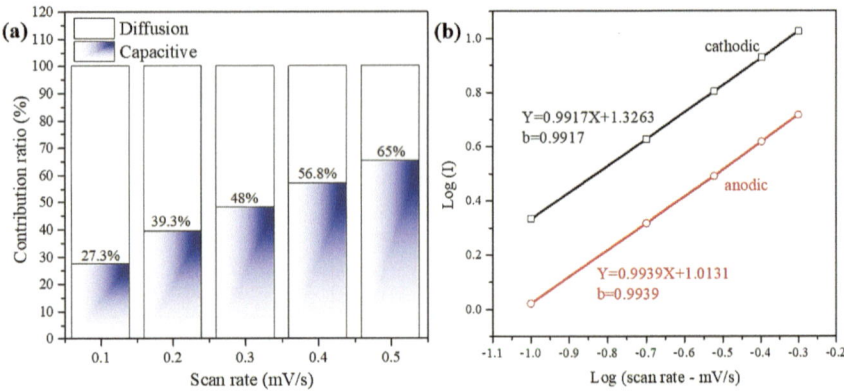

Fig. 4.8   a Ratio of contribution from capacitive and diffusion-controlled electrochemical processes governed by scan rates of 0.1–0.5 mV/s and **b** the plots of log I versus log scan rate in 3 M KOH solution. (Reprinted with permission from [23] Copyright 2023 American chemical society)

Due to the process not being governed by diffusion, the ions present in the solution are adsorbed on the surfaces of electroactive materials. As a result, materials can be processed without pulverization, which is beneficial for the enduring cyclability of SC material [35]. Additionally, in order to attain a high specific capacitance, the capacitive-controlled process benefits from the ability to reduce the mass-transport distance and increase the charge-transfer rate. A relationship between scan rate (V/t) and peak current (i) can be described with $i = a(v/t)^b$, in which a is a constant and b is a slope of the equation [36]. The electrochemical reaction is primarily governed by a diffusion process involving ion insertion or extraction when $b = 0.5$. During the redox reaction, it exhibits a pseudocapacitance behavior if $b = 1$. The slope of the cathodic and anodic peaks during the CV measurements is depicted in Fig. 4.8b as a function of log I versus log scan rate, where the scan rate ranges from 0.1 to 0.5 mV/s. The cathodic and anodic peaks have slopes of 0.9917 and 0.9939, respectively, which indicate that the electrochemical process is mainly characterized by pseudocapacitance behavior.

4.7   Electrochemical Performance of Assembled NVO// Active Carbon Fuel-Cell Supercapacitor

To demonstrate that the NVO-5 cathode is capable of storing and releasing electrons to power the external circuit, a full-cell SC of NVO-5/AC using KOH-PVA gel as the electrolyte was also constructed. The CV curves of an NVO-5/AC cell with varying scan rates and a window potential ranging from 0 to 2.2 V in KOH-PVA gel electrolyte are depicted in Fig. 4.9a. The behavior of pseudocapacitance is represented by a distinct curve characterized by rapid and continuous reversible oxidation processes occurring at the surfaces of the electrodes. In addition, when the scan rate increases, the region of the CV curve for the NVO-5/AC cell expands, indicating that the cell is capable of accumulating charges at a greater potential across the electrodes. The GCD curves observed during charge–discharge procedures at various current densities are illustrated in Fig. 4.9b. The Cs values determined for 1, 3, 5, and 10 A/g are 390, 137, 75, and 10 F/g, respectively, in accordance with the discharge time. Despite the comparatively low Cs values in relation to the half-cell NVO-5, the energy and power densities that have been computed are 286.5 Wh/Kg and 1150 W/Kg, respectively. At 3 M KOH, the energy and power densities are comparatively greater than those in half-cell NVO-5, owing to an enhanced intercalation reaction of K^+ ions into the anode of the AC. Table 4.2 provides comprehensive Cs, Ed, and Pd values for various current densities. Furthermore, the outcomes of the capacitance retention and Coulombic efficiency measurements are demonstrated in Fig. 4.9c, which are derived from GCD cycling at 20 A/g. Following 10 K cycles, the capacitance retention and Coulombic efficiency may reach 97% and 100%, respectively. While the XPS composition analysis revealed that the surface V atoms dissolved into the solution, it is expected that the V atoms would be redeposited as vanadium oxide on the cathode surfaces. Consequently, the electron transport to the surface of the cathode can continue to ensure the cyclability test remains stable. The viable charge storage capabilities of the NVO-5/AC cells that were simply assembled in KOH-PVA gel to power an LED lamp are illustrated in Fig. 4.9d. LED lamps may be illuminated for twenty minutes by NVO-5/AC cells following a full charge, indicating a viable industrial application for NVO-5 material (Fig. 4.10).

Table 4.3 presents a comparison table with comparable SC systems for the purpose of benchmarking against certain performance criteria. In specific capacity and cyclability tests, which are critical indicators in SC technology, the NVO-5 cathode material exhibited superior performance compared to other works.

A particular kinetics may be proposed in elucidating the potential reaction mechanism of the high capacitive NVO-x cathode material, given the observed changes in the chemical states of Ni during the XPS examination subsequent to the 10-K cyclability test. An alteration in the oxidation numbers of Ni from 2^+ to 3^+ was noted, which indicates the presence of a reversible charge balance in the lattice during the discharge process, since $Ni^{3+} + e^- = Ni^{2+}$. Similarly, the charging process will result in the extraction of an electron from the NVO-x electrode, leading to an inverted reaction of $Ni^{2+} = Ni^{3+} + e^-$. The pseudocapacitive nature of the surface reaction on the NVO-x cathode is confirmed by the CV and GCD analyses. Equation (4.5) illustrates that the introduction of electrolyte ions (K^+) into the NVO-x framework will result in a modification of the oxidation number of Ni [52].

$$(NVO - x)_{surface} + yK^+ + xe^- = \left(NVO - xK_y\right)_{surface} \qquad (4.5)$$

Vanadium dissolution in PVA-KOH gel or KOH electrolyte may have a negligible effect on the performance of NVO-x, as confirmed by an ICP result. Despite the confirmation of V solubility in alkaline solutions, the data indicated that the VO_x material retained its long-term cyclability for a total of 10,000 cycles. Equations (4.6–4.8)

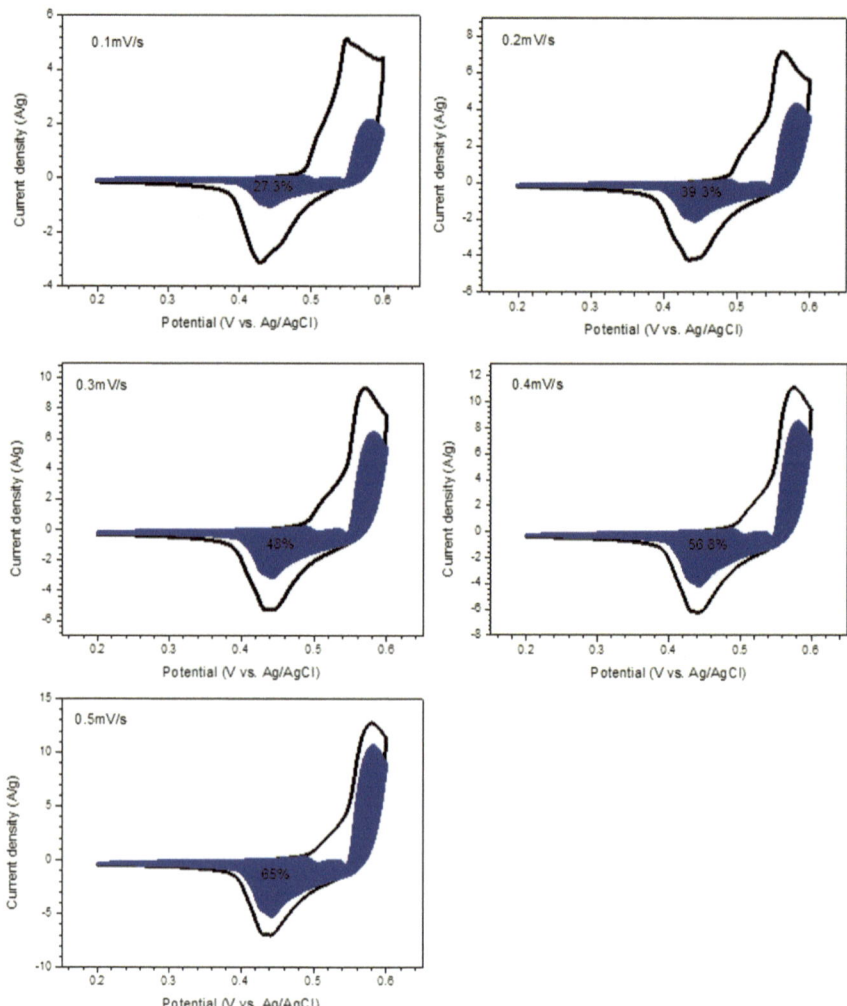

Fig. 4.9 CV spectra of NVO-5 with different scan rates from 0.1 to 0.5 mV/s to indicate the capacitive-controlled processes as shown in blue regions in the spectra. (Reprinted with permission from [23] Copyright 2023 American chemical society)

Table 4.2 Specific Cs, E_d, and P_d values of NVO-5//AC cell in 3 M KOH solution

Current density (A/g)	Specific capacitance (F/g)	Energy density (Wh/Kg)	Power density (W/Kg)
1	390	286.5	1150
3	137	100	3428
5	75	55	5740
10	10	7.5	11,740

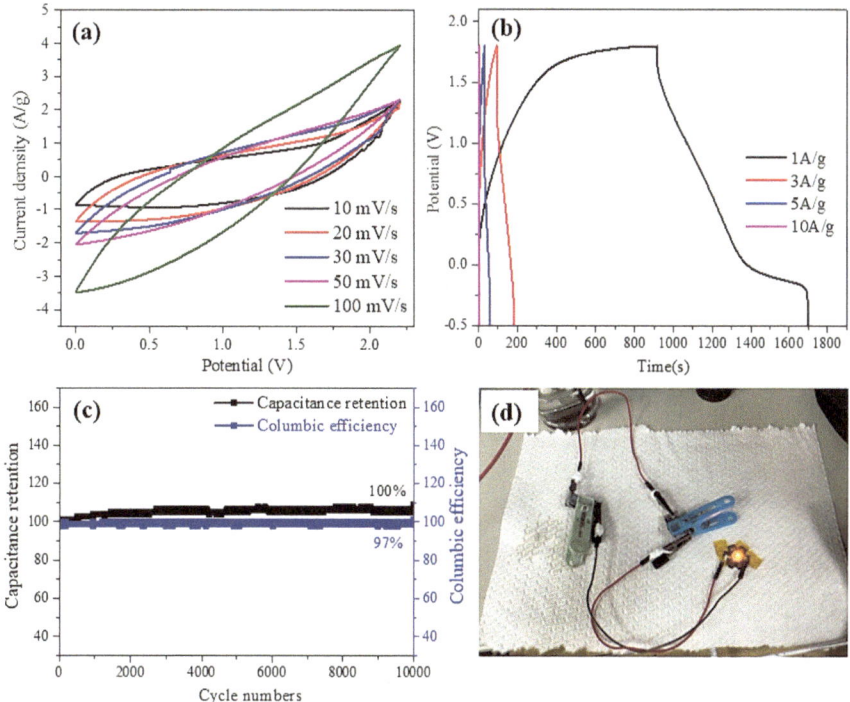

Fig. 4.10 **a** CV, **b** GCD measurement data, **c** capacitance retention and Coulombic efficiency of NVO-5//AC full-cell SC, and **d** an assembled NVO-5//AC full-cell SCs in KOH-PVA gel to light up an LED lamp. (Reprinted with permission from [23] Copyright 2023 American chemical society)

illustrate that this occurrence suggests a potential reversible reaction that may occur during the charge–discharge process. As shown in Eq. (4.6), when oxygen is present in the surrounding environment, water will reduce oxygen to produce a hydrogen peroxide radical in basic solution [53]. Simulating the electro-Fenton-like reaction described in Eqs. (4.7–4.8), the produced hydrogen peroxide and hydroxyl radicals will then react with VO_x on electrodes to produce vanadate ions [54] (Fig. 4.11).

$$O_2 + H_2O + 2e^- = HO_2^- + OH^- \tag{4.6}$$

$$a - VO_x + HO_2^- = VO_4^{3-} + OH + OH^- \tag{4.7}$$

$$a - VO_x + OH + e^- = VO_4^{3-} + H_2O \tag{4.8}$$

Table 4.3 Electrochemical characteristics of NVO-5 cathode as compared to other vanadium-related cathodes

Supercapacitor materials	Electrolyte	Window voltages (V)	Performances (F/g)	Cyclability test (%)	References
NVO-5/Ni foam	3 M KOH	0.4	7500 F/g at 1 A/g	93.1% after 10 K cycles	[23]
NiCo$_2$S$_4$@NiV-LDH/NF	3 M KOH	1.6	1725 F/g at 0.3 A/g	80% after 5 K	[37]
Fe–VOS	0.5 M Na$_2$SO$_4$	0.8	217 F/g at 3 A/g	92% after 4 K cycles	[38]
FeVO$_4$	1 M LiPF$_6$	2.0	220 mAh/g at 0.05 A/g	78% after 40 cycles	[39]
rGO/V$_2$O$_5$	0.5 M K$_2$SO$_4$	1.0	484 F/g at 0.5 A/g	83% after 1 K cycles	[40]
Mo–V–ZnO	2 M KOH	1.6	585 F/g at 0.5 A/g	98% after 2 K cycles	[41]
V$_2$O$_5$–VO$_2$/rGO	1 M Na$_2$SO$_4$	1.7	468.5 F/g at 1 A/g	89.8% after 10 K cycles	[42]
V$_x$Te$_y$/MWCNTs	1 M LiClO$_4$	1.0	470 F/g at 2 mV/s	82.5% after 10 K cycles	[43]
(NH$_4$)$_2$V$_{10}$O$_{25}$·8H$_2$O (NVO)	NH$_4$Cl/PVA	1.8	324 mF/cm^2 at 1 mA/cm^2	71% after 14 K cycles	[44]
rGO@V$_2$O$_5$	5 M LiNO$_3$	0.6	83.5 F/g at 0.5 A/g	73.3% after 1 K cycles	[45]
V$_6$O$_{13}$@C	1 M Na$_2$SO$_4$	0.7	108.9 F/g at 0.2 A/g	34.5% after 1 K	[46]
VS$_2$@Carbon cloth	1 M Na$_2$SO$_4$	0.3	972.5 mF/g at 1 mA/cm^2	95% after 2 K	[47]
V-doped Co$_3$S$_4$	6 M KOH	0.5	3557.6 F/g at 1 A/g	77.5% after 10 K	[48]
V$_2$O$_5$	2 M KOH	1.6	205.06 F/g at 2 mV/s	–	[49]
NCV-LDH NSAs	6 M KOH	0.5	2960 F/g at 1 A/g	81.8% after 10 K cycles	[50]
VN@C	6 M KOH	1.5	17.5 F/g at 1 A/g	88.3% after 2 K	[51]

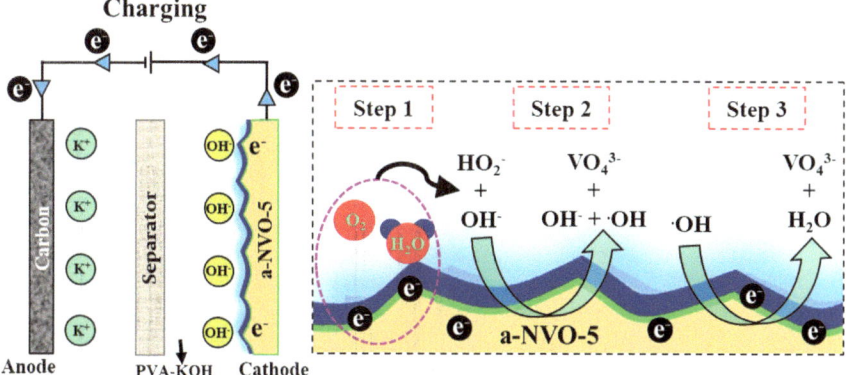

Fig. 4.11 Schematic representation of the charge storage mechanism of the NVO-5 electrode in a 3 M KOH solution. (Reprinted with permission from [23] Copyright 2023 American chemical society)

References

1. J. Tang, Y. Ge, J. Shen, M. Ye, Facile synthesis of $CuCo_2S_4$ as a novel electrode material for ultrahigh supercapacitor performance. Chem. Commun. **52**(7), 1509–1512 (2016)
2. L. Liu, Nano-aggregates of cobalt nickel oxysulfide as a high-performance electrode material for supercapacitors. Nanoscale **5**(23), 11615–11619 (2013)
3. K. Zhang, Z. Cen, F. Yang, K. Xu, Rational construction of $NiCo_2O_4$@Fe_2O_3 core-shell nanowire arrays for high-performance supercapacitors. Prog. Nat. Sci.: Mater. Int. (2021)
4. H. Wan, L. Li, J. Zhang, X. Liu, H. Wang, H. Wang, Nickel Nanowire@Porous $NiCo_2O_4$ Nanorods arrays grown on Nickel foam as efficient pseudocapacitor electrode. Front. Energy Res. **5**(33) (2017)
5. P. Lu, X. Jiang, W. Guo, L. Wang, T. Zhang, Y. Boyjoo, W. Si, F. Hou, J. Liu, S.X. Dou, J. Liang, A Ni–Co sulfide nanosheet/carbon nanotube hybrid film for high-energy and high-power flexible supercapacitors. Carbon (2021)
6. Y.-Z. Su, K. Xiao, N. Li, Z.-Q. Liu, S.-Z. Qiao, Amorphous $Ni(OH)_2$ @three-dimensional Ni core–shell nanostructures for high capacitance pseudocapacitors and asymmetric supercapacitors. J. Mater. Chem. A **2**(34), 13845–13853 (2014)
7. L. Wan, C. He, D. Chen, J. Liu, Y. Zhang, C. Du, M. Xie, J. Chen, In situ grown NiFeP@$NiCo_2S_4$ nanosheet arrays on carbon cloth for asymmetric supercapacitors. Chem. Eng. J. **399**, 125778 (2020)
8. Y. Miao, X. Zhang, J. Zhan, Y. Sui, J. Qi, F. Wei, Q. Meng, Y. He, Y. Ren, Z. Zhan, Z. Sun, Hierarchical NiS@CoS with controllable core-shell structure by two-step strategy for supercapacitor electrodes. Adv. Mater. Interfaces **7**(3), 1901618 (2020)
9. S. Guan, X. Fu, Z. Lao, C. Jin, Z. Peng, NiS-MoS_2 hetero-nanosheet arrays on carbon cloth for high-performance flexible hybrid energy storage devices. ACS Sustain. Chem. Eng. **7**(13), 11672–11681 (2019)
10. Y. Yan, B. Li, W. Guo, H. Pang, H. Xue, Vanadium based materials as electrode materials for high performance supercapacitors. J. Power. Sources **329**, 148–169 (2016)
11. H. Qin, S. Liang, L. Chen, Y. Li, Z. Luo, S. Chen, Recent advances in vanadium-based nanomaterials and their composites for supercapacitors. Sustain. Energy Fuels **4**(10), 4902–4933 (2020)

12. X. Wang, B. Shi, X. Wang, J. Gao, C. Zhang, Z. Yang, H. Xie, One-step synthesis of V_2O_5/Ni_3S_2 nanoflakes for high electrochemical performance. J. Mater. Chem. A **5**(45), 23543–23549 (2017)

13. J. Balamurugan, G. Karthikeyan, T.D. Thanh, N.H. Kim, J.H. Lee, Facile synthesis of vanadium nitride/nitrogen-doped graphene composite as stable high performance anode materials for supercapacitors. J. Power. Sources **308**, 149–157 (2016)

14. W. Sun, G. Gao, Y. Du, K. Zhang, G. Wu, A facile strategy for fabricating hierarchical nanocomposites of V_2O_5 nanowire arrays on a three-dimensional N-doped graphene aerogel with a synergistic effect for supercapacitors. J. Mater. Chem. A **6**(21), 9938–9947 (2018)

15. I. Shakir, Z. Ali, J. Bae, J. Park, D.J. Kang, Layer by layer assembly of ultrathin V_2O_5 anchored MWCNTs and graphene on textile fabrics for fabrication of high energy density flexible supercapacitor electrodes. Nanoscale **6**(8), 4125–4130 (2014)

16. D.P. Dubal, O. Ayyad, V. Ruiz, P. Gómez-Romero, Hybrid energy storage: the merging of battery and supercapacitor chemistries. Chem. Soc. Rev. **44**(7), 1777–1790 (2015)

17. T. Schoetz, L.W. Gordon, S. Ivanov, A. Bund, D. Mandler, R.J. Messinger, Disentangling faradaic, pseudocapacitive, and capacitive charge storage: a tutorial for the characterization of batteries, supercapacitors, and hybrid systems. Electrochim. Acta **412**, 140072 (2022)

18. S. Wen, Y. Liu, F. Zhu, R. Shao, W. Xu, Hierarchical MoS_2 nanowires/$NiCo_2O_4$ nanosheets supported on Ni foam for high-performance asymmetric supercapacitors. Appl. Surf. Sci. **428**, 616–622 (2018)

19. M.S. Vidhya, G. Ravi, R. Yuvakkumar, D. Velauthapillai, M. Thambidurai, C. Dang, B. Saravanakumar, Nickel–cobalt hydroxide: a positive electrode for supercapacitor applications. RSC Adv. **10**(33), 19410–19418 (2020)

20. H.B. Li, Y.Q. Gao, G.W. Yang, Electrochemical route for accessing amorphous mixed-metal hydroxide nanospheres and magnetism. RSC Adv. **5**(56), 45359–45367 (2015)

21. B. Grégoire, C. Ruby, C. Carteret, Hydrolysis of mixed Ni^{2+}–Fe^{3+} and Mg^{2+}–Fe^{3+} solutions and mechanism of formation of layered double hydroxides. Dalton Trans. **42**(44), 15687–15698 (2013)

22. P. Shvets, O. Dikaya, K. Maksimova, A. Goikhman, A review of Raman spectroscopy of vanadium oxides. J. Raman Spectrosc. **50**(8), 1226–1244 (2019)

23. H. Abdullah, S.-J. Jhuang, H. Shuwanto, D.-H. Kuo, High Charge storage of amorphous Ni-doped VOx-Modified $Ni(OH)_2$ substrate on a Ni foam cathode in a base solution. ACS Appl. Energy Mater. **6**(2), 898–909 (2023)

24. L. Li, J. Xu, J. Lei, J. Zhang, F. McLarnon, Z. Wei, N. Li, F. Pan, A one-step, cost-effective green method to in situ fabricate $Ni(OH)_2$ hexagonal platelets on Ni foam as binder-free supercapacitor electrode materials. J. Mater. Chem. A **3**(5), 1953–1960 (2015)

25. D. Xiong, W. Li, L. Liu, Vertically aligned porous nickel(II) hydroxide nanosheets supported on carbon paper with long-term oxygen evolution performance. Chem.—Asian J. **12**(5), 543–551 (2017)

26. K. Kishi, K. Fujiwara, Ultrathin nickel oxide on the $V_2O_3Cu(100)$ surface studied by XPS. J. Electron Spectrosc. Relat. Phenom. **71**(1), 51–59 (1995)

27. H. Abdullah, N.S. Gultom, D.-H. Kuo, Depletion-zone size control of p-type NiO/n-type Zn(O, S) nanodiodes on high-surface-area SiO_2 nanoparticles as a strategy to significantly enhance hydrogen evolution rate. Appl. Catal. B **261**, 118223 (2020)

28. H. Abdullah, D.-H. Kuo, X. Chen, High efficient noble metal free Zn(O, S) nanoparticles for hydrogen evolution. Int. J. Hydrogen Energy **42**(9), 5638–5648 (2017)

29. M. Shahid, J. Liu, Z. Ali, I. Shakir, M.F. Warsi, Structural and electrochemical properties of single crystalline MoV_2O_8 nanowires for energy storage devices. J. Power. Sources **230**, 277–281 (2013)

30. D. Chen, H. Tan, X. Rui, Q. Zhang, Y. Feng, H. Geng, C. Li, S. Huang, Y. Yu, Oxyvanite V_3O_5: a new intercalation-type anode for lithium-ion battery. InfoMat **1**(2), 251–259 (2019)

31. C.V.V.M. Gopi, R. Vinodh, S. Sambasivam, I.M. Obaidat, S. Singh, H.-J. Kim, Co_9S_8–$Ni_3S_2/CuMn_2O_4$–$NiMn_2O_4$ and $MnFe_2O_4$–$ZnFe_2O_4$/graphene as binder-free cathode and anode materials for high energy density supercapacitors. Chem. Eng. J. **381**, 122640 (2020)

32. D.P. Chatterjee, A.K. Nandi, A review on the recent advances in hybrid supercapacitors. J. Mater. Chem. A **9**(29), 15880–15918 (2021)

33. F. Zhang, J. Ma, H. Yao, Ultrathin Ni–MOF nanosheet coated $NiCo_2O_4$ nanowire arrays as a high-performance binder-free electrode for flexible hybrid supercapacitors. Ceram. Int. **45**(18, Part A), 24279–24287 (2019)

34. X. Zhang, J. Wang, Y. Sui, F. Wei, J. Qi, Q. Meng, Y. He, D. Zhuang, Hierarchical nickel-cobalt phosphide/phosphate/carbon nanosheets for high-performance supercapacitors. ACS Appl. Nano Mater. **3**(12), 11945–11954 (2020)

35. M.S. Rahmanifar, H. Hesari, A. Noori, M.Y. Masoomi, A. Morsali, M.F. Mousavi, A dual Ni/Co–MOF–reduced graphene oxide nanocomposite as a high performance supercapacitor electrode material. Electrochim. Acta **275**, 76–86 (2018)

36. X. Zhang, M. Jin, Y. Zhao, Z. Bai, C. Wu, Z. Zhu, H. Wu, J. Zhou, J. Li, X. Pan, E. Xie, Improved lithium-ion battery performance by introducing oxygen-containing functional groups by plasma treatment. Nanotechnology **32**(27), 275401 (2021)

37. E. Niknam, H. Naffakh-Moosavy, S.E. Moosavifard, M. Ghahraman Afshar, Amorphous V-doped Co_3S_4 yolk-shell hollow spheres derived from metal-organic framework for high-performance asymmetric supercapacitors. J. Alloy. Compd. **895**, 162720 (2022)

38. P. Asen, S. Shahrokhian, A.I. Zad, Iron-vanadium oxysulfide nanostructures as novel electrode materials for supercapacitor applications. J. Electroanal. Chem. **818**, 157–167 (2018)

39. Y. Si, G. Liu, C. Deng, W. Liu, H. Li, L. Tang, Facile synthesis and electrochemical properties of amorphous $FeVO_4$ as cathode materials for lithium secondary batteries. J. Electroanal. Chem. **787**, 19–23 (2017)

40. B. Saravanakumar, K.K. Purushothaman, G. Muralidharan, Fabrication of two-dimensional reduced graphene oxide supported V_2O_5 networks and their application in supercapacitors. Mater. Chem. Phys. **170**, 266–275 (2016)

41. M.R. Pallavolu, J. Nallapureddy, R.R. Nallapureddy, G. Neelima, A.K. Yedluri, T.K. Mandal, B. Pejjai, S.W. Joo, Self-assembled and highly faceted growth of Mo and V doped ZnO nanoflowers for high-performance supercapacitors. J. Alloy. Compd. **886**, 161234 (2021)

42. Y. Chen, P. Lian, J. Feng, Y. Liu, L. Wang, J. Liu, X. Shi, Tailoring defective vanadium pentoxide/reduced graphene oxide electrodes for all-vanadium-oxide asymmetric supercapacitors. Chem. Eng. J. **429**, 132274 (2022)

43. B. Pandit, S.R. Rondiya, R.W. Cross, N.Y. Dzade, B.R. Sankapal, Vanadium telluride nanoparticles on MWCNTs prepared by successive ionic layer adsorption and reaction for solid-state supercapacitor. Chem. Eng. J. **429**, 132505 (2022)

44. P. Wang, Y. Zhang, H. Jiang, X. Dong, C. Meng, Ammonium vanadium oxide framework with stable NH^{4+} aqueous storage for flexible quasi-solid-state supercapacitor. Chem. Eng. J. **427**, 131548 (2022)

45. H. Liu, W. Zhu, D. Long, J. Zhu, G. Pezzotti, Porous V_2O_5 nanorods/reduced graphene oxide composites for high performance symmetric supercapacitors. Appl. Surf. Sci. **478**, 383–392 (2019)

46. W. Yang, J. Zeng, Z. Xue, T. Ma, J. Chen, N. Li, H. Zou, S. Chen, Synthesis of vanadium oxide nanorods coated with carbon nanoshell for a high-performance supercapacitor. Ionics **26**(2), 961–970 (2020)

47. M.Y. Zhang, J.Y. Miao, X.H. Yan, Y.H. Zhu, Y.L. Li, W.J. Zhang, W. Zhu, J.M. Pan, M.S. Javed, S. Hussain, Vanadium disulfide nanosheets loaded on carbon cloth as electrode for flexible quasi-solid-state asymmetric supercapacitors: energy storage mechanism and electrochemical performance. J. Mater. Chem. C **10**(2), 640–648 (2022)

48. J. Pan, S. Li, F. Li, W. Zhang, D. Guo, L. Zhang, D. Zhang, H. Pan, Y. Zhang, Y. Ruan, Design and construction of core-shell heterostructure of Ni–V layered double hydroxide composite electrode materials for high-performance hybrid supercapacitor and L-tryptophan sensor. J. Alloy. Compd. **890**, 161781 (2022)

49. D. Velpula, S. Konda, S. Vasukula, S.C. Chidurala, Microwave radiated comparative growths of vanadium pentoxide nanostructures by green and chemical routes for energy storage applications. Mater. Today: Proc. **47**, 1760–1766 (2021)

50. Z. Wu, D. Khalafallah, C. Teng, X. Wang, Q. Zou, J. Chen, M. Zhi, Z. Hong, Vanadium doped hierarchical porous nickel-cobalt layered double hydroxides nanosheet arrays for high-performance supercapacitor. J. Alloy. Compd. **838**, 155604 (2020)

51. M.R. Pallavolu, Y. Anil Kumar, R.R. Nallapureddy, H.R. Goli, A. Narayan Banerjee, S.W. Joo, In-situ design of porous vanadium nitride@carbon nanobelts: a promising material for high-performance asymmetric supercapacitors. Appl. Surf. Sci. **575**, 151734 (2022)

52. J.S. Daubert, N.P. Lewis, H.N. Gotsch, J.Z. Mundy, D.N. Monroe, E.C. Dickey, M.D. Losego, G.N. Parsons, Effect of meso- and micro-porosity in carbon electrodes on atomic layer deposition of pseudocapacitive V$_2$O$_5$ for high performance supercapacitors. Chem. Mater. **27**(19), 6524–6534 (2015)

53. Y. Xue, Y. Wang, S. Zheng, Z. Sun, Y. Zhang, W. Jin, Efficient oxidative dissolution of V$_2$O$_3$ by the in situ electro-generated reactive oxygen species on N-doped carbon felt electrodes. Electrochim. Acta **226**, 140–147 (2017)

54. S. Chen, J. Hong, H. Yang, J. Yang, Adsorption of uranium (VI) from aqueous solution using a novel graphene oxide-activated carbon felt composite. J. Environ. Radioact. **126**, 253–258 (2013)

Chapter 5
Advances of Vanadium Oxides in Supercapacitor Applications

5.1 Promising Transition Metal Oxides (TMOs) in Supercapacitor Applications

Transition metal oxides (TMO) are among the pseudocapacitive materials that are sought after as electrode materials for pseudocapacitors. This is primarily attributable to the unique properties of TMOs, including their intrinsically high stability, changeable valence that enables ions and electrons to intercalate into the lattice of metallic compounds, and flawless pseudocapacitance [1, 2]. Ruthenium oxide (RuO_2) has been the subject of the most scholarly investigation among the TMOs owing to its exceptional electrical conductivity and high specific capacitance (up to $1580\,Fg^{-1}$), which make it one of the most appealing possibilities for supercapacitors [1, 3]. However, the scarcity, high cost, and acute toxicity of ruthenium have restricted its extensive range of potential applications [4, 3, 5, 6]. Further investigation has been conducted on metal oxides such as NiO, Co_3O_4, MnO_2, and V_2O_5, which have been demonstrated to be abundant, economically viable, and environmentally benign [1], in addition to RuO_2. A multitude of metal oxide electrodes are characterized by inadequate electrical conductivity [7, 8]. Ternary metal oxides, denoted as AB_2O_4 (where A or B represents Ni, Co, Mo, Mn, and so forth), are distinguished from binary metal oxides by their greater electrical conductivity and active reaction sites, which are the results of the element synergistic effects and the coexistence of two metal ions [7, 9]. Vanadium oxides and adjacent compounds have garnered growing interest in recent times due to their unique chemical and physical characteristics, as well as their peculiar layered structure. As cathode materials for reversible lithium batteries [10, 11] and SCs [12], in addition to functioning as catalysts [13], intelligent thermochromic windows, and gas sensors [14], these V compounds have a wide range of applications. Vanadium oxides consist of a variety of binary oxides denoted by the formula VO_x^{2+}. These oxides also encompass the oxidation states V_2O_5, V_6O_{13}, V_3O_7, VO_2, and others. Nevertheless, these oxides exhibit subpar electrical conductivity and cycling properties. In order to enhance the conductivity

© The Author(s), under exclusive license to Springer Nature Singapore Pte Ltd. 2024   77
H. Abdullah, *Vanadium Oxide-Based Cathode for Supercapacitor Applications*,
SpringerBriefs in Applied Sciences and Technology,
https://doi.org/10.1007/978-981-97-5243-0_5

and durability of electrodes based on VO, several modifications were implemented, including doping, nanocomposite formation and structural, morphological, and electrical property adjustments [15–17]. Systematic investigation has been conducted in the crystal structure of vanadium oxides, the design principle for optimizing the electrochemical performance of VO-based materials synthesis techniques, and the physical, chemical, and morphological properties of V_2O_5 and VO_2-based materials. Alternatively, the effects of the additions and modifications on the performance of the various parameters are also discussed. It is noteworthy that synthesis approaches have been discovered to play a critical role in improving the electrochemical performance of materials composed of vanadium oxide. The investigation of high-performance materials that are suitable for supercapacitors is facilitated by morphology. Thus, many kinds of vanadium oxide-based materials are highlighted in detail alongside their features. In conclusion, prospects and obstacles are examined to illuminate how advancements in these material categories might augment the practicality of supercapacitors.

As vanadium has the electronic configuration $[Ar]4s^2 \, 3d^3$, the most common valences of vanadium are $+5$, $+4$, $+3$, and $+2$, whereas its oxidation state can range from $+5$ to -3 [18]. Four vanadium oxides have a single oxidation state, although several others have mixed oxidation states ($+2$ for VO, $+3$ for V_2O_3, $+4$ for VO_2, and $+5$ for V_2O_5). The oxidation states of $+5$ (orange to yellow), $+4$ (blue), and $+3$ (green) are the oxidation states that cause color variation [19]. The formation of octahedral, pentagonal bipyramid, square pyramid, and tetrahedral structures with shared corners, edges, or faces is caused by the oxygen coordination in the crystalline structures of vanadium oxides [18]. Solid V–O bonds comprise the layered structure of V_2O_5. A vanadium atom positioned at the apex of a distorted square pyramid composed of five oxygen atoms forms five V–O bonds, which are situated at the apex of the layered V_2O_5 crystal structure. Furthermore, a common vertex and a co-edge are utilized to connect the square pyramids in order to produce a planar framework. In reality, the phase equilibria of the vanadium oxygen system are highly complex; altering the valence of vanadium will result in a substantial structural change. To illustrate, the VO_2 framework bears resemblance to tunnels and is composed of twisted VO_6 octahedra that share edges and corners [20]. Two layers of V_2O_5 are depicted in Fig. 5.1a from a perspective along the b-axis and c-axis, while the various phases of VO_2 and their corresponding lattice characteristics are illustrated in Fig. 5.1b. By combining and permuting polyhedral coordination, it is possible to generate a multitude of vanadium oxides featuring diverse open frameworks, such as tunnel-like and layer structures. As a result of the ability of multivalent oxides to boost electrochemical activity and conductivity [21, 22], mixed-valence vanadium oxide is also gaining significant attention. The physicochemical properties of the distinct phases of vanadium oxides are substantially altered because of the oxidation state of the vanadium cations.

Charge is predominantly stored in an electrical energy storage system through the accumulation of charge on the electrode surface or by means of electron gain and loss followed by ion insert/extract. These mechanisms are in favor of electrode materials characterized by high conductivity and specific surface area [23]. In general,

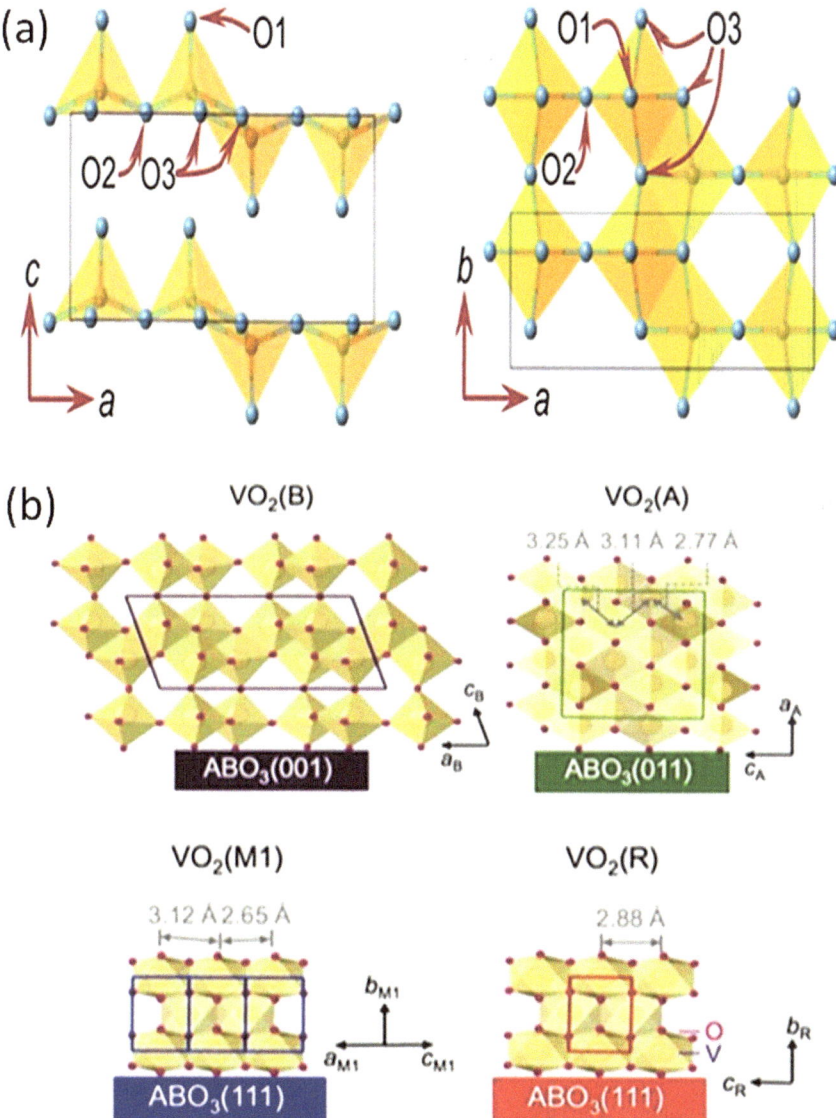

Fig. 5.1 **a** Structure of two layer V_2O_5 along the b-axis and c-axis with gray balls as the V atoms and blue balls as O atoms and **b** VO_2 with lattice parameters [19]. (Reproduced with permission under CC-BY 4.0)

the fundamental factor can be attributed to the distinctive characteristics of transition metal oxides (TMOs), including their notable inherent stability and challengeable valence, which is a pseudocapacitance charge that allows ions and electrons to inter-calate into the metallic compound lattice [24–27]. It is critical to have the ability to consistently generate and refine V-compounds in order to fulfill certain demands for energy storage device applications and get the desired performance attributes.

5.2 Design of VO_x-Based Materials

5.2.1 Surface Morphology

The charge storage mechanism is influenced by the diffusion process due to surface shape. The aberrant behavior shown by nanostructures can be elucidated by tuning Fick's law ($t \pm L^2/D$), which posits that a reduction in diffusion length (L) offers a shorter ion diffusion time (t). This occurrence lends credence to the notion that nanostructuring facilitates a higher influx of ions over the interface, hence accelerating the response of vanadium-based electrodes. When the diameters of V compounds approach the nanoscale, their pseudocapacitive characteristics become more pronounced. This characteristic is beneficial in addressing the slow solid-state diffusion of intercalating ions [28]. The reaction taking place in electrodes with nanostructures effectively inhibits the intercalation-induced phase transition, resulting in superior kinetics compared to the conventional diffusion-controlled Faradaic reaction. This is due to the ability of a substantial number of charges to accu-mulate in close proximity to the surface within timescales ranging from microseconds to milliseconds. It can be established a connection between the previously indistin-guishable rate capability and the emerging capacitive controlled storage mechanism in nanostructured V-compounds based on this presumption [29].

5.2.2 Tuning of Electronic Structures

According to kinetics studies, the lattice environment significantly influences how cations interact with host materials. Modification can be accomplished with remark-able success through doping. Ex situ or in situ methods are employed to introduce metal ions or nonmetallic heteroatoms into the electrode material structure. The electrochemical performance of electrode materials can be enhanced through the introduction of doping ions or atoms, which alters their electronic distribution [30]. The correction for the electrical neutrality imbalance in the crystal lattice is achieved through the doping of cations, which introduces defects such as holes and vacancies [31]. Consequently, electronic conductivity may be enhanced by the development of holes. The relationship between ion diffusivity and activation energy is represented by the classic Arrhenius equation: $D_i = D_0 \exp(-\Delta G/K_B.T)$; where ΔG represents the energy barrier, k_B denotes the Boltzmann constant, T signifies the temperature, and D_0 does not account for the preexponential factor [30]. Potentially the most effective method for modifying the activation energy of V-compounds is via lattice engineering via ion doping and lattice expansion. However, in contrast, cations and micromolecules have the ability to function as pillars in order to expand the interlayer space, preserve the structure, and increase the capacity. Conversely, the electrochemical environment between the V–O layers can be altered through intercalation of cations, which influences the arrangement of electron clouds within the V–O layer. The introduction of foreign atoms into vanadium oxides can result in an enhancement of their electrical and ionic conductivities. At this moment, alkali metal ions, transition metal ions, and alkaline earth metal ions are suitable cations for intercalation. Variations in the quantity of ions introduced, their valence state, and ion radius induce structural modifications in the intercalated products. The reason is during the intercalating process, cations with discrete ion radii induce unique deformations in a single layer [32, 33]. Furthermore, structural alterations are influenced by the variance in binding energies between oxygen atoms and cations in the V–O layer. It is important to note, however, that metal ions with large ionic radii and high molar masses, such as Ag and Cu, can occupy a considerable amount of interlayer space. Suboptimal energy densities and particular capabilities may ensue as a consequence [23]. Heteroatoms that are not metallic are employed in an extra-efficient doping technique. Frequently, elements including nitrogen, phosphorus, sulfur, or sulfur dioxide are doped into carbon-based materials or conductive polymers that are mixed with vanadium oxide-based materials during the non-metallic heteroatom doping process. Doping the electrode with elements that possess a strong electronegative valence significantly enhances its cycle stability and rate performance [34]. The higher bonding force between the composite materials is the major effect to stability. In a recent year, it has seen the addition of conductive polymers to the interlayer of vanadium oxide-based materials, including polyaniline (PANI), polypyrrole (PPy), and poly(3,4-ethylenedioxythiophene) (PEDOT) [35]. Polymer-V exhibited remarkable rate capability and capacity retention [36, 37]. By incorporating a conductive polymer into the electrode, cycle performances and rates are enhanced by virtue of the polymer's capacity to facilitate the passage of electrons and ions.

5.2.3 Optimization of Synergistic Effects

Through the regulation of chemical reactivity and surface chemistry, surface engineering improves conductivity and structural integrity. Conformal carbon coating is widely used for V-compounds due to its ability to offer several advantages in an easy manner through the utilization of carbon sources that are easily obtainable, including polymers, polysaccharides, and organic acids [38]. The enhanced electrochemical capabilities are due to an increase in electronic and ionic conductivity, and mechanical confinement, which mitigates excessive volume growth. The implementation of a synergistic coupling effect through the juxtaposition of various material types presents a viable solution to the sluggish kinetics of electrical and ionic transport encountered by the majority of V-compounds. In order to maximize the potential of V-compounds, carbon-based scaffolds can provide them with exceptionally suited qualities, among other possible combinations [29]. For instance, the enhancement of electrochemical properties in carbonaceous materials resulting from the incorporation of V_2O_5 [39], $Li_3V_2(PO_4)_3$ [31], and $Na_3V_2(PO_4)_3$ [40]. This improvement can be attributed to the subsequent factors: Carbon supports with high conductivity facilitate efficient charge transfer, hence averting significant agglomeration and safeguarding the structural integrity of V-compounds via carbon confinement. Furthermore, carbon has a remarkable level of tunability, enabling it to transform into an extensive array of shapes with the introduction of defects and vacancies caused by doped or absent atoms. The investigation of conductive scaffolds comprised of organic materials with intrinsic doping (e.g., polymers and ionic liquids), chemically generated doped frameworks, and materials derived from living organisms has been spurred by these discoveries. The carbon that has been changed has enhanced wettability on its surface and cation-friendly binding sites. The enhanced reaction kinetics may be attributed to these unique characteristics, in addition to the synergistic enhancement of the electrode's electronic conductivity facilitated by carbon materials [41].

5.3 Different VO$_x$ Electrode Materials

Vanadium oxides (VO_x) are a class of transition metal oxides that are becoming more and more popular in the energy industry due to their varied structural variations, high theoretical specific capacity, and wide output voltage range [42]. However, the electrical conductivity and cycling properties of these oxides are weak. Optimizing electrical conductivity, structural flexibility, bandgap, and charge carrier mobility is essential to solve issues with the electrochemical charge storage mechanism of the supercapacitor. Thus, the most effective solutions to these issues are heterostructure nanomaterials and structural alterations [43–45]. Creating nano heterostructures is the best method for utilizing fast reversible faradic reactions in addition to near-surface ion adsorption for energy storage. The creation of active sites and defects in the grain boundaries of heterostructure materials results in multiple redox activity, increased ionic conductivity, and a short diffusion path. Multi-metal heterostructures are also commonly used as supercapacitor electrodes and have a variety of crystalline structures. By creating multi-metal oxides of transition metals, it is possible to combine the superior electrochemical characteristics of two or more metal oxides [46]. However, in doped electrode materials, the energy levels of the host and guest in the highest occupied molecular orbital (HOMO) and lowest unoccupied molecular orbital (LUMO) can be precisely adjusted to control the electrical properties of poorly conductive materials, leading to the desired properties of high-power density, fast charge/discharge, and long life [46]. By combining EDLC with pseudocapacitive materials, one can further improve the electrochemical properties of the materials by increasing their surface area and active electrochemical sites [47].

5.3.1 Pristine V$_2$O$_5$ in Different Unique Morphologies

Previous studies have demonstrated that vanadium oxide capacitive performances could be enhanced by unique designs because of the charge storage mechanism. Furthermore, compared to bulk materials, electrode materials with distinctive morphologies frequently have higher energy densities [48, 49]. The electrical, optical, magnetic, and catalytic properties of materials, as well as their prospective uses in contemporary materials synthesis, are greatly influenced by their morphologies, microstructures, and organizational patterns. These properties provide us with a feasible way to use shape control to synthesize function-specific materials [50, 51]. Many attempts were undertaken to obtain a high surface area and short diffusion distance in order to produce V$_2$O$_5$ with different morphologies for high-performance electrode materials [52, 53]. Despite considerable advancements in recent decades, shape-controlled synthesis of nano- and microstructures remains a challenging task of great significance. For example, it has been demonstrated that hydrothermal synthesis is a highly effective and useful technique for creating inorganic materials with clearly defined shape, size, and structure. It is carried out in a sealed aqueous solution and does not require the use of specialized equipment or stringent experimental conditions like high temperatures [54, 55]. The most often employed methods for structural alterations on V$_2$O$_5$ are (i) modifying the hydrothermal solvent usage, (ii) enhancing the surface area through preheating activation that increases the transport of electrons and ions during the charge/discharge process, (iii) differing parameters for synthesis, (iv) adjusting the amounts of material, (v) varying kinds and quantities of acid and solvent use. Therefore, developing materials with distinctive structures to improve their properties remains challenging and crucial for material scientists [49]. Miao Li et al. were able to produce V$_2$O$_5$ nanosheets with different morphologies by modifying the solvent under a straightforward hydrothermal process, based on the theory of improving the electrochemical performance of V$_2$O$_5$ powders by modifying the structure and shape. For optimal process conditions, the water-to-ethanol volume ratio should be 4:1. Electrochemical test results show that the V$_2$O$_5$ nanosheet functions effectively in electrolytes with a concentration of 0.5 M K$_2$SO$_4$. The specific capacitance may reach 375 F.g^{-1} at a current density of 0.5 A.g^{-1} and after 1000 cycles, the specific capacitance retention rate reaches 96.7%. [56]. Additionally, Chunyu Peng et al. employed a scalable and environmentally benign method to produce V$_2$O$_5$ nanobelts using commercial powder type V$_2$O$_5$ as precursors. The CV curves demonstrate that the powder type V$_2$O$_5$−1 M Na has superior electrochemical characteristics when compared to V$_2$O$_5$, V$_2$O$_5$–0.5 M Na, and V$_2$O$_5$−2 M

Na. At a bending angle of 120°, $V_2O_5 - 1$ M Na solid-state supercapacitors maintain 90.7% of their specific capacitance, exhibiting an energy density of 13.9 Wh kg^{-1} and a power density of 1250 W kg^{-1} [57].

The hydrothermal method was utilized by Balamuralitharan B. et al. to produce a V_2O_5 nanorod electrode material. The electrode in 0.5 M Na$_2$SO$_4$ electrolyte demonstrated good long-term stability (80% over 5000 cycles) and a high areal specific capacitance of 417.3 mF cm^{-2}. This was in contrast to the electrode in 0.5 M H$_2$SO$_4$ electrolyte. Notable characteristics included very low impedance values, high energy density, longer-term stability, and higher specific capacitance. The exceptional capacitive performance was primarily caused by the nanorod shape [58]. V_2O_5 nanorods were made by Swapnil S. Karade et al. via a straightforward two-step procedure that included preheating activation and a hydrothermal method. The one-dimensional structure enabled the use of a large surface area with suitable access for electron/ion transfer during the charge/discharge process. The as-prepared V_2O_5 electrode showed good energy storage with a high specific capacitance of 347 F.g^{-1} at 1 A.g^{-1} and remarkable electrochemical stability of 91% after 10,000 CV cycles. A coin cell setup was also used to evaluate the two-electrode performance. Interestingly, we observed a high specific capacitance of 172 F.g^{-1} at 1.9 A.g^{-1}. The V_2O_5 symmetric supercapacitor coin cell also demonstrated a maximum energy density of 23.9 Wh kg^{-1} and a power density of 937 W kg^{-1}. The device stability of 94.3% after 10,000 cycles indicates that the V_2O_5 electrode is promising for real-world use [59]. Jiqi Zheng et al. crystallized an aqueous solution of ammonium metavanadate (NH$_4$VO$_3$) at 8 °C and then calcined the resulting porous V_2O_5 flakes in an environmentally benign and straightforward process. The resulting V_2O_5 flakes had an average pore size of 28.66 nm, which is predominantly composed of mesoporous pores. The findings of a study on the effects of temperature showed that the product with the biggest capacitance was the one calcined at 400 °C. The good result is due to its right crystalline and pore structure, observed in SEM analysis in Fig. 5.2. As a battery-type electrode, porous V_2O_5 flakes demonstrated a high capacity of up to 510 F.g^{-1} at 0.2 A.g^{-1}, as well as perfect cycling stability with >110% retention after 1000 cycles [52].

Using commercial V_2O_5 and H$_2$C$_2$O$_4$·2H$_2$O as raw materials, Yifu Zhang et al. combined a mix of calcination and a simple hydrothermal technique to synthesize three distinct types of V_2O_5 structures. The morphology of the sample is impacted by the concentration of oxalic acid used in the research. V_2O_5 nanobelts, nanoparticles, and microspheres were created using 0.63, 1.89, and 3.78 g of oxalic acid, respectively, as shown in Fig. 5.3. V_2O_5 microspheres have the highest specific capacitance of 308 F.g^{-1} when used as supercapacitor electrodes in 1 mol L^{-1} LiNO$_3$ electrolyte. CV curves that serve as the example are provided in Fig. 5.4. The specific capacitances of all V_2O_5 electrodes may be diminishing with more cycles due to the electrode material dissolving in aqueous LiNO$_3$ solution [60]. To produce hollow, spherical, high specific surface area V_2O_5 particles, spray pyrolysis is a fairly simple and scalable method.

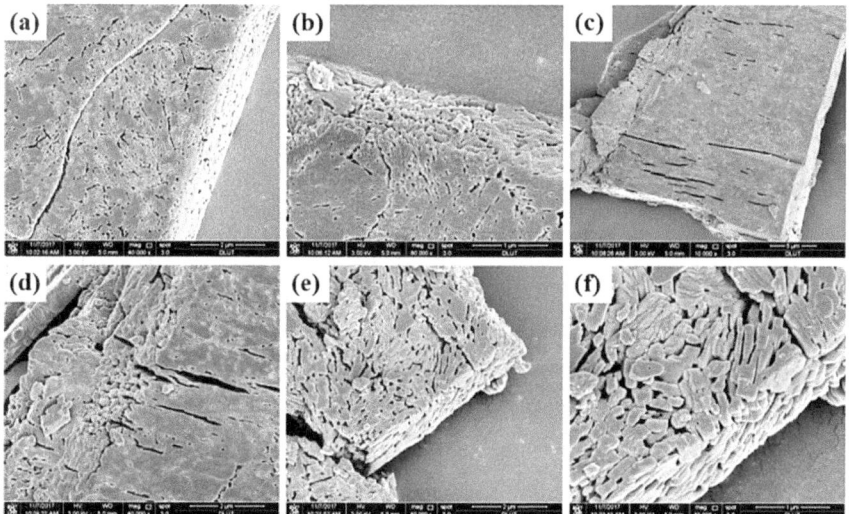

Fig. 5.2 SEM images of V$_2$O$_5$ flakes calcined at different temperatures of **a, b** 300 °C, **c, d** 400 °C, **e, f** 500 °C for 4 h [52] (Elsevier Copyright 2018 with license number: 5761770559290)

Additionally, it offers inherent advantages for microsphere structural control and stoichiometry over conventional approaches. As-prepared microsphere structure, shape, and electrochemical performance are all substantially influenced by the calcination temperature. The variation in electrical performance caused by a shift in calcination temperature is seen in Fig. 5.5 [61]. Based on the concept, hollow V$_2$O$_5$ microspheres were fabricated by Zhendong Yin et al. via a straightforward spray pyrolysis procedure. The V$_2$O$_5$ microspheres composed of regular nanorods with smooth surfaces were generated from the micron-sized, smooth-surfaced, hollow, and spherical particles after a 3-h post-treatment process at 500 °C in a nitrogen environment. V$_2$O$_5$ particles that had been post-treated exhibited increased surface roughness, enhanced crystallinity, and resistance to aggregation. As a consequence, its electrochemical performance in supercapacitors is improved. During the assessment of the impact of electrolyte concentration on performance, it was observed that 5 M of LiNO$_3$ performed admirably within the concentration range of 1 to 5 M. The variation in electrical performance caused by a change in electrolyte content is seen in Fig. 5.6 [62]. By employing the solvothermal template approach followed by calcination, Y. Zhang synthesized hollow V$_2$O$_5$ microspheres from NH$_4$VO$_3$, ethylene glycol, and carbon spheres as the initial components. At current densities of 0.5 A.g^{-1}, 1 A.g^{-1}, 2 A.g^{-1}, and 5 A.g^{-1}, hollow V$_2$O$_5$ microspheres exhibited favorable pseudocapacitance characteristics, with specific capacitances of 488 F.g^{-1}, 455 F.g^{-1}, 434 F.g^{-1}, and 396 F.g^{-1}, respectively. In addition to exhibiting a power density of 900 W kg^{-1}, they displayed an extraordinary energy density of 8.784 × 10^5 J kg^{-1} [63].

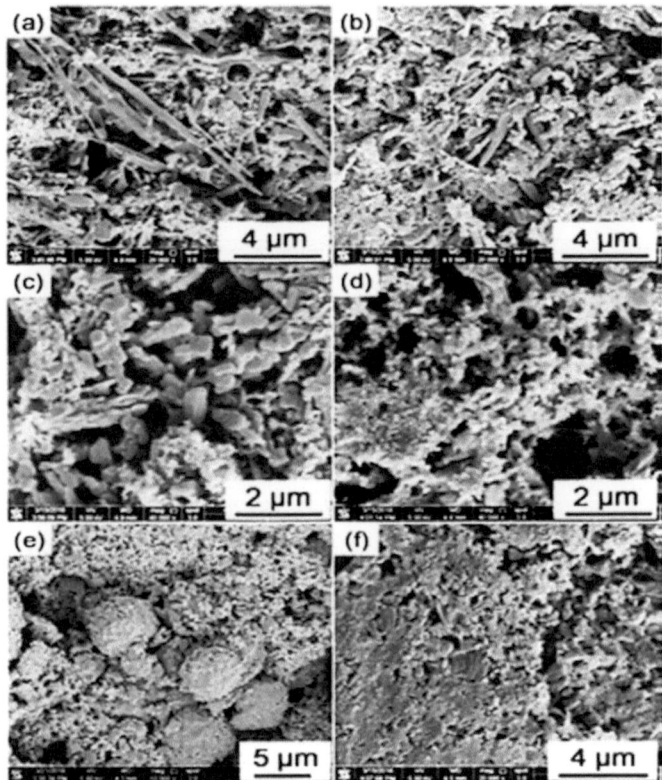

Fig. 5.3 SEM images of V_2O_5 electrodes **a, c, e** before and **b, d, f** after the cycles with the shapes of **a, b** nanobelts, **c, d** nanoparticles, and **e, f** microspheres [60] (Elsevier Copyright 2016 with license number: 5763421357521)

Juyi Mu and colleagues designed nanoflowers, nanoballs, nanowires, and nanorods, which are all examples of typical hydrothermal morphologies of pure V_2O_5 nanostructures, as shown in Fig. 5.7. The particle morphology is influenced by the type of solvent and acid used. Zero-dimensional nanoballs and hierarchical nanoflowers composed of V_2O_5 nanocrystals were generated using C_2H_5OH as the solvent. One-dimensional nanowires and nanorods were generated by supplementing the procedure with H_2O. As shown in Fig. 5.8, CV curves, the storage capacity, electrochemical kinetics, and rate capabilities are all greatly improved by the rod-like V_2O_5, as determined by electrochemical tests. One-dimensional V_2O_5 nanorods demonstrate the highest specific capacitance of 235 Fg^{-1} at a current density of 1 Ag^{-1} when employed as a supercapacitor electrode in a 1 mol/L Na_2SO_4 electrolyte [65]. V_2O_5 structures self-assemble by means of quadrate sheets, resembling "multilayer cakes," in contrast to the geometries previously described. Layer-by-layer vanadium oxide quadrate structures were produced using a straightforward hydrothermal method. In addition, the electrochemical performance of the V_2O_5

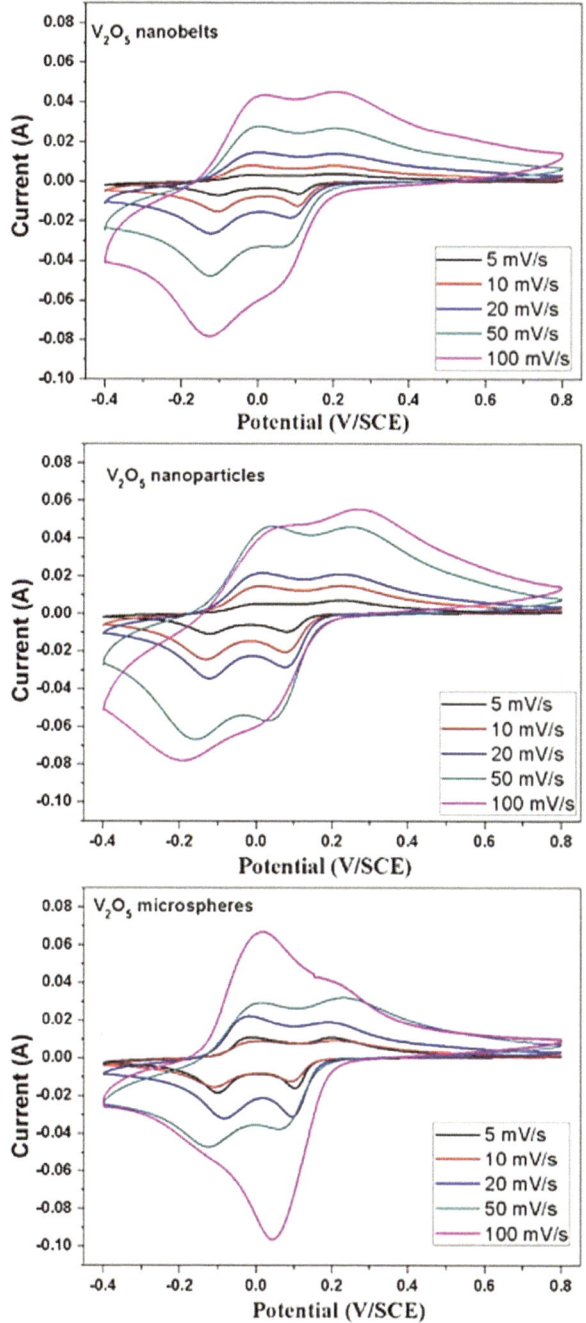

Fig. 5.4 CV curves of different V$_2$O$_5$ electrodes with various scan rates [60] (Elsevier Copyright 2016 with license number: 5772220012956)

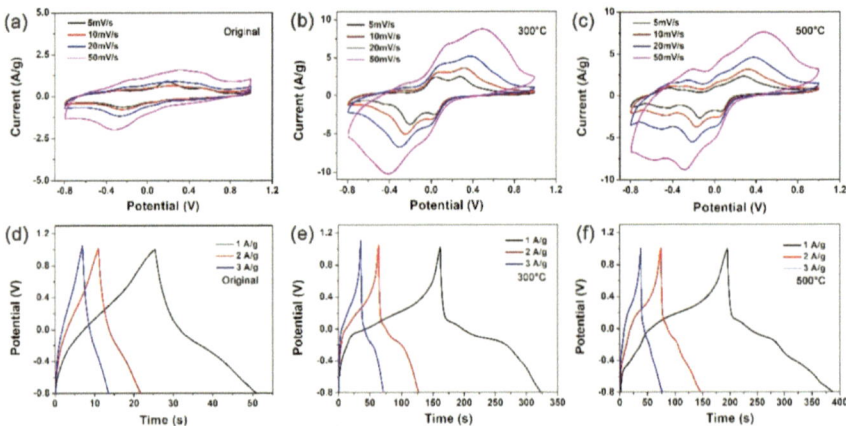

Fig. 5.5 CV curves of V₂O₅ microspheres after treatment at different temperatures, **a** original, **b** 300 °C, **c** 500 °C in 1 M LiNO₃ electrolyte, with the related **d, e, f** GCD curves [64] (Elsevier Copyright 2024 with license number: 5765150914354)

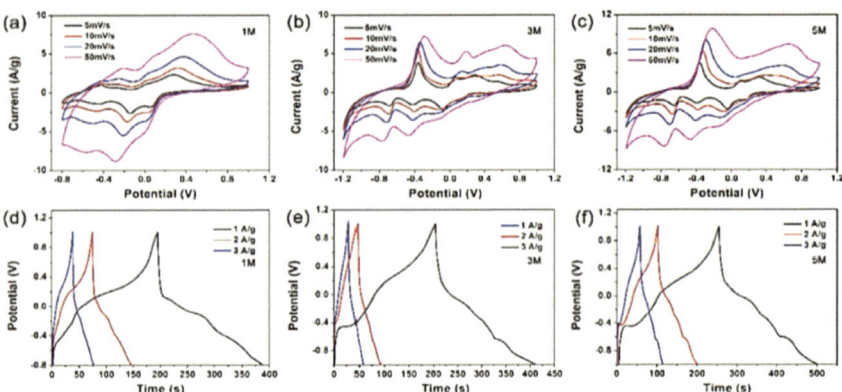

Fig. 5.6 CV curves of 500 °C-treated V₂O₅ microspheres in **a** 1 M, **b** 3 M, **c** 5 M LiNO₃ and **d, e, f** their corresponding GCD curves [64] (Elsevier Copyright 2024 with license number: 5765160292775)

electrode is inferior [49, 60, 66] due to the dissolution of V during the charging/discharging process, a common occurrence for VO_x electrodes utilized in SCs and Li-ion batteries. Yifu Zhang et al. resolved this issue by employing propylene carbonate as the organic electrolyte. It was shown that this approach substantially improved the electrochemical properties of V_2O_5 quadrate structures composed of layers upon layers. The electrochemical properties of the layer-by-layer V_2O_5 structures as electrodes in the SCs device were assessed utilizing aqueous and organic electrolytes, CV, and GCD. In an organic electrolyte, the specific capacitance at 1 A.g^{-1} is 347 F.g^{-1}, which is 46% greater than the value in an aqueous electrolyte (238 F.g^{-1}). A hundred cycles during cycle performance reduce the specific capacitances of V$_2$O$_5$ layer-by-layer structures to approximately 30% and 82%, respectively, of the initial discharge capacity in aqueous and organic electrolytes. Increasing the cycle number does not significantly affect the specific capacitance of an organic electrolyte beyond 40 cycles [49].

In order to demonstrate how varying parameters during sample preparation can impact the physical and chemical performance of V$_2$O$_5$ samples, Meenal D. Patil et al. synthesized V$_2$O$_5$ powder via the thermal decomposition method and examined the effects of distinct annealing temperatures (V1, V2, and V3, representing 400 °C, 500 °C, and 600 °C, respectively) on the physicochemical and electrochemical properties of the powder. The active V$_2$O$_5$–Ni foam electrode exhibited pseudocapacitive qualities as evidenced by its electrochemical characteristics. Based on the CV curves presented in Fig. 5.9, the V1 electrode has the largest specific capacitance of 1227.2 F g^{-1} compared to the V2 (723.8 F g^{-1}) and V3 (511.2 F g^{-1}) electrodes using a three-electrode setup and a scan rate of 1 mV s^{-1}. Despite undergoing 2000 consecutive GCD cycles, the perfect electrode retained an approximate 96.1% specific capacitance [67]. In contrast to 1-D or 2-D carbon electrodes, titanium mesh possesses a very complex 3-D architecture that considerably enhances the capacitance of SCs to accept various charges by providing numerous contact areas and movement points. This makes the entire concept scalable by encouraging the eventual application of V$_2$O$_5$. In light of this, Muhammad Sufyan Javed et al. directly fabricate 2-D nanoflakes of V$_2$O$_5$ on a titanium substrate as a novel electrode for a pseudocapacitor. The electrode possesses remarkable charge storage efficiency, consistent structural stability, and effective electrical/ionic transport. With a remarkable rate capability, V$_2$O$_5$@Ti possesses a capacitance of 1520 F g^{-1} at 1.5 Ag^{-1} and a consistent cycling stability of 99% for a duration of 12,000 cycles [68]. Utilizing a straightforward, affordable, additive-free, and highly reproducible hydrothermal method, K. K. Purushothaman et al. effectively manufactured V$_2$O$_5$ nanoflowers on a nickel foam backbone. In the absence of supplementary components such as carbon black and polymer binder, the synthesized material can be evaluated in its current state as electrode material for supercapacitors. The sample exhibits enhanced rate capacity, improved cycle stability, and increased specific capacitance due to a good material morphology. The aforementioned substance exhibits enhanced electrochemical characteristics, such as an elevated capacitance of 601 F.g^{-1}, an increased ability to sustain higher current densities of 373 F.g^{-1} at 5 A.g^{-1}, and enhanced cycle stability. Most significantly, the symmetric supercapacitor with 3D-V$_2$O$_5$ at the Ni foam electrode has an energy density of 29 Wh Kg^{-1} and a specific capacitance of 142 F.g^{-1}.

The best types of electrolytes are aqueous ones since they have higher charge storage capacities than other types due to their high electrical conductivity and ion content [69–71]. The electrochemical properties of the V electrode in an aqueous electrolyte with 1 M KOH are shown in Fig. 5.9. A comparison is made between the potential scan rates of 1 mV s^{-1} for raw Ni-foam and electrodes manufactured at different annealing temperatures. The images of V1–NF, V2–NF, and V3–NF are displayed at scan rates of 1–100 mV s^{-1}. The electrode material produced a

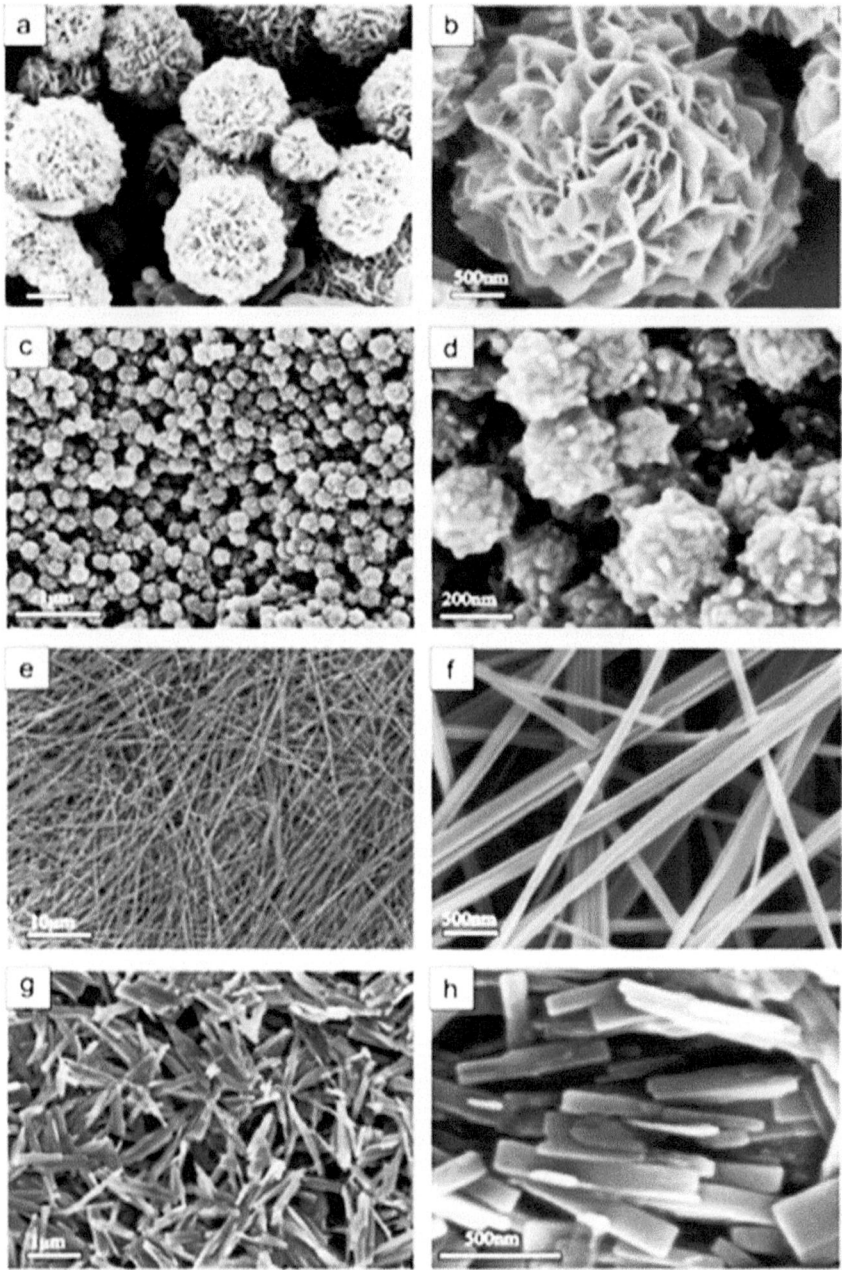

Fig. 5.7   SEM images of the **a, b** nanoflower, **c, d** nanoball, **e, f** nanowire, and **g, h** nanorod V$_2$O$_5$ [65] (Elsevier Copyright 2015 with license number: 5766271256794)

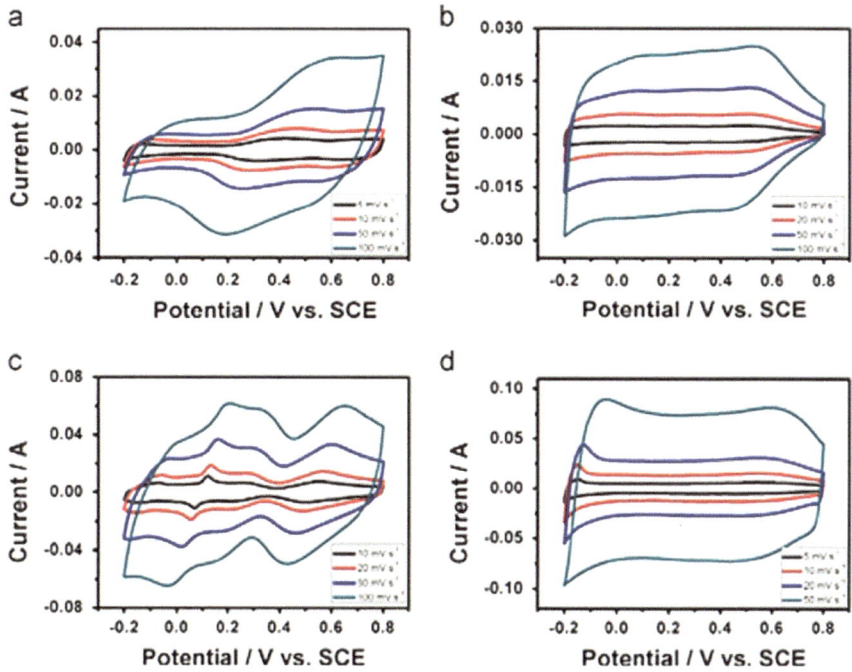

Fig. 5.8 CV curves of different nanostructures of **a** nanoflower, **b** nanoball, **c** nanowire, and **d** nanorod V$_2$O$_5$ electrode in Na$_2$SO$_4$ electrolyte with different scan rates [65]. (Elsevier Copyright 2015 with license number: 5766271474967)

noteworthy specific capacitance of 403 F.g^{-1} at a current density of 1 A.g^{-1} in a mixed electrolyte solution of 0.5 M KOH and 1 M Na$_2$SO$_4$. In this mixed electrolyte, the V$_2$O$_5$ electrode showed good cyclic stability for 3000 cycles [72], with an 85% capacity retention at 10 A.g^{-1}. The work covers the current developments in the structure, morphology, and electrochemical performance of V$_2$O$_5$, as well as the efforts of other researchers to employ this material as an electrode material in supercapacitor applications.

5.3.2 V$_2$O$_5$–Metal Oxide Composites

Mitigating the disadvantages of V$_2$O$_5$ and Fe$_3$O$_4$ nanoparticles is the most desirable option because of their higher energy storage capacity, natural richness, low cost, and abundance as a metal oxide on Earth. Fe$_3$O$_4$ nanoparticles placed on the surface of V$_2$O$_5$ significantly increase the electroactive area and involvement in the electrochemical reaction. Bhargav Akkinepally et al. successfully fabricate Fe$_3$O$_4$ nanoparticles using a straightforward method, as well as bundled V$_2$O$_5$ nanobelts that are adorned with Fe$_3$O$_4$

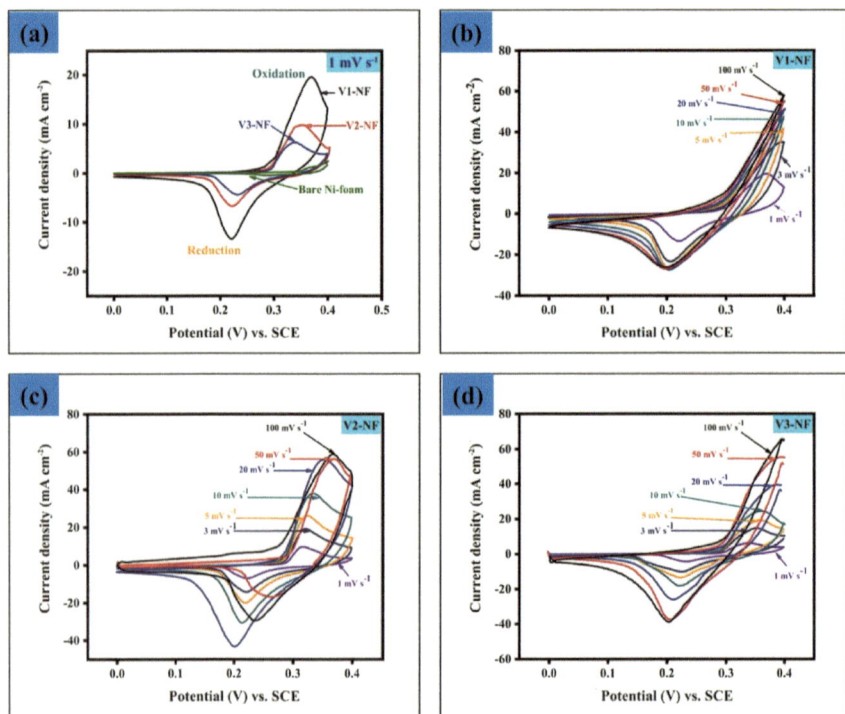

Fig. 5.9 Electrochemical activities of V electrodes in 1 M KOH aqueous electrolyte by comparing **a** CV plots of V electrodes fabricated at different annealing temperatures and bare Ni-foam with a potential scan rate of 1 mV s^{-1}, **b** V1–NF, **c** V2–NF, **d** V3–NF at various scan rates of 1–100 mV s^{-1} [67]. (Elsevier Copyright 2021 with license number: 5766280872398)

nanoparticles (VF). When comparing the electrochemical parameters of the electrode to those of the Fe$_3$O$_4$ (660.3 F.g^{-1}) and V$_2$O$_5$ (750.1 F.g^{-1}) electrodes, the former attained a high specific capacitance of 1519 F.g^{-1} at 15 A.g^{-1}. This could be a result of the surface-active sites on the modified electrodes being expanded, which greatly enhanced the storage capacity. The electrode cycling stability over 4000 GCD cycles revealed an 84% capacity retention. A built device with VF nanostructures displayed superior GCD kinetics, with a specific capacitance of 93 F.g^{-1} at 0.5 A.g^{-1} and high energy and power densities of 13–34 Wh kg^{-1} and 765–1530 W kg^{-1} respectively, spanning a current density range of 0.5–1 A.g^{-1}. It was observed that 84% of the capacitance of the device was retained after 5000 cycles, demonstrating remarkable long-term stability. The electrical evaluation of the three-electrode system is illustrated in Fig. 5.10 [73]. TiO$_2$/V$_2$O$_5$ mixtures are the most exciting substances to consider for electrodes in supercapacitors. The substitution of metal–metal bonds with Ti–O–V bonds has the potential to enhance chemical stability and electrochemical activity due to the reduced resistance of intra-particle electron hopping. Additionally, degradation and cycle stability-related parameters should be enhanced. The TiO$_2$–V$_2$O$_5$ nanocomposite is a highly favorable choice owing to its economical price, non-toxic nature, high capacitance, and straightforward nanostructure synthesis. By employing a wet chemical methodology, Apurba Ray et al. effectively synthesized interconnecting tube-like structured structures of 3D-TiO$_2$–V$_2$O$_5$. The 1 M KCl solution is utilized to achieve a maximum specific capacitance of 310 F.g^{-1} at a scan rate of 2 mV/s and a molar ratio of Ti:V = 10: 20, as shown in Fig. 5.11. It has been determined that the active interior sites of the electrode, and not the outside surface, are responsible for the highest capacitance value, and that an increase in the V ratio significantly improves the electrochemical behavior of this composite. A significant number of active sites can be provided by the tube-shaped mesoporous nanostructure, which possesses a comparatively high specific surface area [74]. This facilitates the efficient passage of ions.

MnO$_2$ possesses a diverse array of practical implementations, such as supercapacitors, owing to its exceptional thermal and chemical stability, high theoretical specific capacitance, ample natural availability, environmentally benign nature, and economical cost. K. Latha et al. accomplished this by employing a straightforward

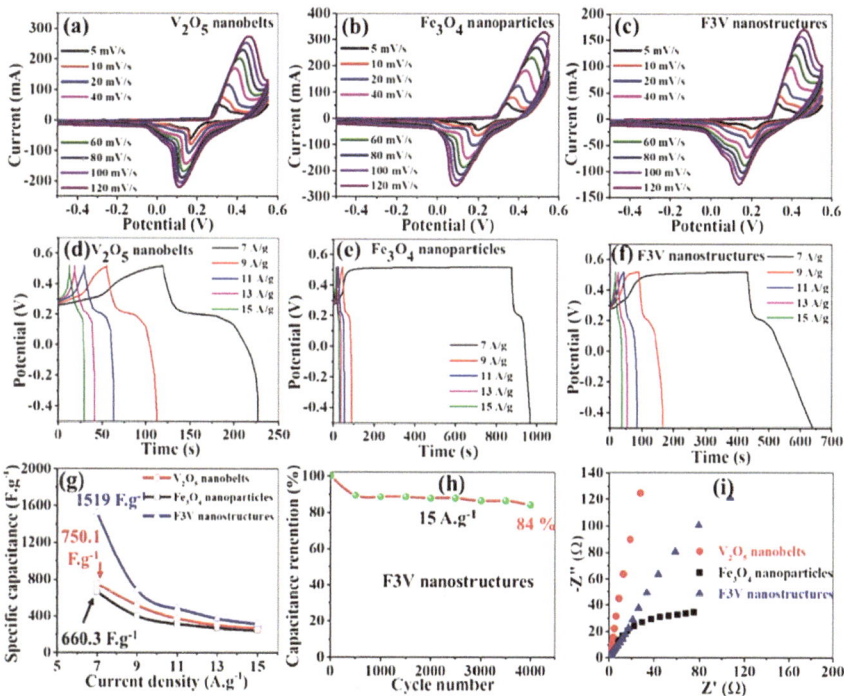

Fig. 5.10 Electrochemical analyses of CV spectra of **a** V$_2$O$_5$ nanobelts, **b** Fe$_3$O$_4$, and **c** VF$_3$ nanostructures at different scan rates as well as GCD spectra of **d** V$_2$O$_5$ nanobelts, **e** Fe$_3$O$_4$, and **f** VF$_3$ nanostructures at various current densities, with **g** the specific capacitance versus current density, **h** capacity retention over 4000 cycles at 15 A.g^{-1}, and **i** electrochemical impedance spectra [73]. (Elsevier Copyright 2023 with license number: 5767990153398)

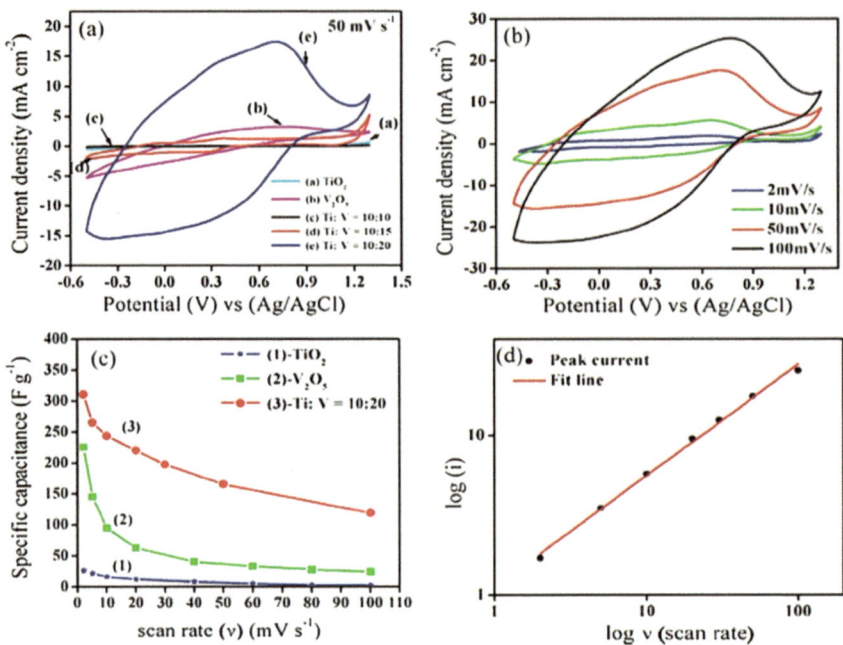

Fig. 5.11 **a** CV spectra of TiO$_2$, V$_2$O$_5$ and TiO$_2$–V$_2$O$_5$ with different concentrations at scan rate of 50 mV/s, **b** CV at different scan rates of TiO$_2$–V$_2$O$_5$ with Ti: V = 10: 20, **c** specific capacitance vs. scan rate of TiO$_2$, V$_2$O$_5$ and its composite (Ti: V = 10: 20) and **d** the plot of peak current vs log of scan rate for Ti: V = 10:20 [74]. (Elsevier Copyright 2018 with license number: 5770030559676)

microwave-assisted technique to produce the MnO$_2$/V$_2$O$_5$ nanocomposites and pure V$_2$O$_5$. With the addition of 6% MnO$_2$ by weight, the electrochemical performance of the V$_2$O$_5$ composite increased from 95 to 269 F.g^{-1} in the H$_2$SO$_4$ electrolyte and from 150 to 570 F.g^{-1} in the KOH electrolyte [74].

5.3.3 Mo and Ni Metal-Incorporated V$_2$O$_5$

Doping refers to the process of augmenting the properties of a host material through the introduction of supplementary components. To improve the electronic conductivity of vanadium pentoxide, it can be doped with manganese [75], nickel [76, 77], copper [78], tantalum [79], or other elements. The electrochemical characteristics of metal oxides, polymers, and carbonaceous materials can be significantly improved through doping. The existence of heteroatoms results in the generation of free electrons, the formation of electrochemically active sites, and an expansion of the electrode material surface area. Doping can boost the rate capacity and stability of pseudo-capacitance materials [46]. In addition, Mo is a suitable metal for doping because of its state of 6+ oxidation, in contrast to Vanadium's stable oxidation state

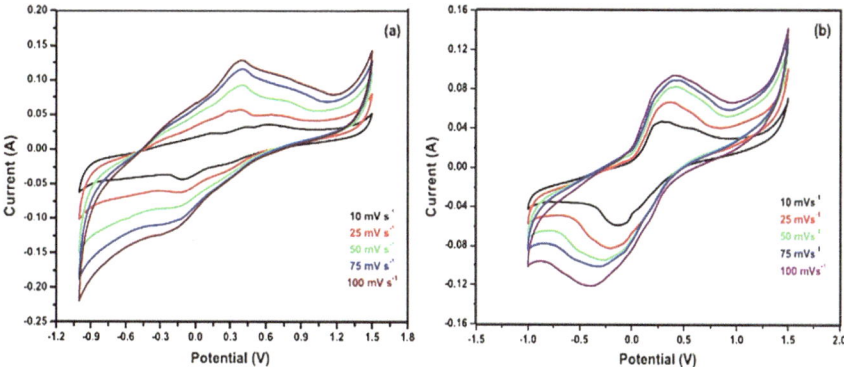

Fig. 5.12 CV curves of V$_2$O$_5$ electrodes **a** with and **b** without Ni doping [81]. (Elsevier Copyright 2022 with license number: 5770030338441)

of V^{5+}. By doping Mo into V$_2$O5, its optical, electrical, and electrochemical characteristics can be altered, resulting in the formation of donor-like defects. Thermal evaporation was employed by N. Guru Prakash et al. to fabricate thin films of Mo-doped V$_2$O$_5$ on nickel substrates at 250 °C and 2%, 4%, or 6% Mo concentration. The greatest specific capacitance of the V$_2$O$_5$ thin films doped with 4% Mo was 175 mF/cm^2 at a current density of 1 mA/cm^2 [80]. In order to demonstrate the impact of Ni doping on the electrochemical characteristics of V$_2$O$_5$, V. Uma Shankar et al. prepared Ni-doped V$_2$O$_5$ nanorods via the sol–gel method and documented pseudo-capacitance behavior in both electrodes (V$_2$O$_5$ and Ni-doped V$_2$O$_5$). However, the specific capacitance of the Ni:V$_2$O$_5$ electrode was 152 F.g^{-1}, which was considerably greater than that of the electrode composed entirely of V$_2$O$_5$. The CV curves of V$_2$O$_5$ and Ni:V$_2$O$_5$ electrodes are depicted in Fig. 5.12. The inclusion of Ni ions improved the structural stability, electrical conductivity, and electrochemical performance of the V$_2$O$_5$ nanorods, as demonstrated by the result [81].

5.3.4   Carbon-Based Materials/V$_2$O$_5$ Composites

In battery technologies, the electrical and electrochemical performance of TMOs and carbon-based materials as anode materials is improved. It is critical to take into account several factors while developing and constructing electrodes: the rate at which charge transfers, the degree of interfacial contact between the electrode and electrolyte, and the quantity of ion exposure on the surface [82]. The electronic conductivity of metal-oxide composites is further enhanced by the incorporation of commercially accessible carbon allotropes and their derivatives. It prevents the agglomeration of metal-oxide nanoparticles throughout cycle testing [79, 83, 84]. One feasible approach to enhance the electrochemical capabilities of electrode materials composed of V$_2$O$_5$ is to employ direct binding of nanoscale V$_2$O$_5$ to an electrically conductive backbone, such as carbon materials [85, 86]. Notwithstanding the considerable amount of research dedicated to V_2O_5, several challenges persist. These include reversible morphological transports caused by ion intercalation and de-intercalation, which lead to volumetric expansion and often result in the detachment of electrode material attachments to the current collector. These effects ultimately compromise the conductivity and stability of the material [87, 88]. The requisite kinetics of ion insertion/de-intercalation are considerably impeded by the polarization effects that inevitably occur at the interfaces between the electrode and electrolyte. Another instance of low ionic diffusivity that necessitates further improvement [89]. In order to optimize operation, hybrid capacitor technologies integrate the advantageous features of EDLCs and pseudocapacitors. The primary objective of recent developments in SC electrode materials has been to increase their energy density. In order to accomplish this, electrode materials with exceptionally high specific capacitances $(F.g^{-1})$ have been developed, enabling cells to operate at elevated voltages [90]. Adding electrically conductive materials to these V_2O_5 nanocomposites, such as activated nanocarbons, carbon nanotubes/nanofibers, graphene, and conjugated polymer nanostructures, substantially improves electronic transport along the interfaces, in addition to enhancing the material chemical stability, mechanical robustness, and redox-active surface areas [91]. Such improvements may lead to the development of more intelligent devices. Carbon-based nanoparticles are favorable for the fabrication of composites [92–94] on account of their increased surface area, improved chemical, mechanical, and adaption resilience, and superior adaptability in comparison to other pseudocapacitive systems. Numerous studies have been devoted to the preparation of carbon-based composites from V_2O_5.

Numerous functional groups present on the GO may potentially operate as valid redox reaction sites for the attachment of various metal oxide ions to the surface. The metal ions may enhance the electrochemical performance of the composite electrodes and the complimentary effects of graphene oxide [76]. Subsequent to its notable attributes of substantial specific surface area, high thermal and electrical conductivity, exceptional mechanical properties, and superior chemical stability. Graphene, a honeycomb lattice-structured two-dimensional (2D) carbon nanomaterial, has been the subject of extensive investigation across various disciplines [85, 95]. The synthetic microwave approach was employed by Min Fu et al. to fabricate composites of V_2O_5 and graphene. A multitude of V_2O_5/graphene sites were synthesized, and their electrochemical properties were examined in depth. The electrochemical performance was optimized with the V_2O_5/graphene-0.341 ratio, as illustrated in Fig. 5.13 [96].

5.3.5 V₂O₅/Conducting Polymer Composites

Recently, there has been considerable interest in materials composed of conducting polymers and transition metal oxides, as these materials combine the benefits of both constituents while retaining their great features. Composites minimize particle size, diminish particle agglomeration, enhance cycle stability, introduce additional pseudocapacitance, and augment surface area. Conductive polymer, by virtue of

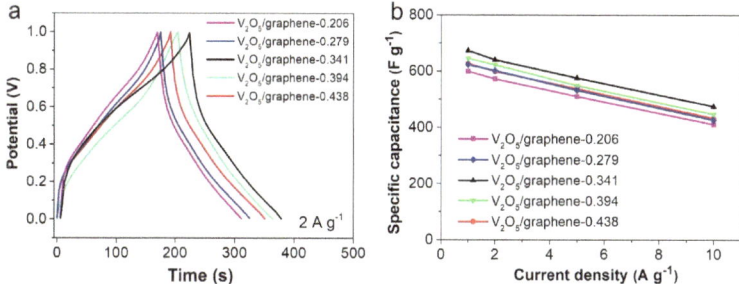

Fig. 5.13 a GCD and **b** Cs with different current densities of different V$_2$O$_5$/graphene composite-based supercapacitors [96] (Elsevier Copyright 2021 with license number: 5770060039104)

its flexible and polymeric properties, imparts excellent electrical conductivity. In contrast, metal oxide serves as a charge storage medium and furnishes conducting polymer with structural and mechanical support. Thus, it is thought that the performance of composite forms is much superior to that of their constituent materials [97]. A "core–shell" architecture, which consists of conducting polymer and active components, has recently been demonstrated to effectively improve the stability of electrodes. For the core–shell structure, which provides the electrode with enhanced electrical conductivity, shorter ion transport paths, and increased cycling stability, the capabilities of each component must first be combined. Secondly, the coating layer can inhibit the aggregation of active elements and their detrimental reactions by reducing the surface energy. PPy-coated V$_2$O$_5$ nanocomposites are produced by Yue Liang et al. via an eco-friendly, cost-effective, and straightforward sol–gel method. In addition to preventing the dissolution of V$_2$O$_5$ in aqueous solutions, the PPy coating layer promotes charge transfer through its high conductivity. The two-electrode symmetric V$_2$O$_5$/PPy device exhibited the maximum energy density of 37 Wh/Kg when the power density was 161 W/kg. Electro-conducting polymer PANI has emerged as a prominent electrode material for pseudocapacitors in recent years. The material has several advantages, such as exceptional thermal stability, notable utility in energy storage technology, highly reversible charge storage capabilities [98], straightforward production, and a high capacitance of 3407 F.g^{-1} [99]. An electrodeposition approach was employed by Asma Aamir et al. to fabricate a high-performance V$_2$O$_5$–PANI composite, which was subsequently coated onto a metallic nickel foam substrate to serve as an electrode for pseudocapacitors. Ni foam, with its porous and conductive properties, effectively reduces the length of the ion diffusion channel. For charge storage, the V$_2$O$_5$–PANI composite exhibited the greatest specific capacitance of 1115 F.g^{-1} at a current density of 1 A.g^{-1} and an exceptionally broad voltage window of 2.5 V [97]. By utilizing an electrodeposition technique, Ming-Hua Bai et al. produce a vanadium oxide and polyaniline (PANI) composite that functions as a high-performance negative electrode. The VP electrode had a wide potential window of 1.6 V, which facilitated charge storage. At

$0.5\,\text{mA cm}^{-2}$, it demonstrated the highest capacitance of $443\,\text{F.g}^{-1}$ or $664.5\,\text{mF cm}^{-2}$ [100].

Considering that conducting polymers, metal oxides, and graphene are the three most important electrode materials for supercapacitors, a logical and persuasive combination of these three substances would produce innovative electrode materials with substantially improved electrochemical results. As a result, the merger of graphene with pseudocapacitive substances (including conductive polymers and transition metal oxides) can significantly enhance the capacitance of carbon materials and the stability of the pseudocapacitive materials during charge–discharge cycles. An original ternary nanocomposites, comprising vanadium pentoxide (V_2O_5), polypyrrole (PPy), and graphene oxide (GO), are synthesized via a unidirectional electrochemical deposition process. As seen in Fig. 5.14, the electrochemical deposition of V_2O_5/PPy/GO nanocomposite involves the electrochemical deposition of PPy, GO, and V_2O_5 onto a stainless steel (SS) substrate via an aqueous solution including vanadyl acetate, pyrrole, and GO. The nanocomposite formed from V_2O_5/PPy/GO showcases a remarkable specific capacitance of $750\,\text{F.g}^{-1}$ at $5\,\text{A.g}^{-1}$, indicating that the desired increase in capacitance value may be achieved by integrating the EDLC of GO with the pseudocapacitance characteristics of PPy and V_2O_5 (Fig. 5.15). Furthermore, after undergoing 3-K cycles of charge–discharge experiments, the 2D and 3D AFM pictures present in Fig. 5.16, the V_2O_5/PPy/GO/SS electrode retains 83% of its initial specific capacitance value. At an applied current density of $5\,\text{A.g}^{-1}$, the symmetric V_2O_5/PPy/GO device exhibits a maximum energy density of 27.6 Wh/kg and a high-power density of 13,680 W/kg [101]. Juncy Parayangattil Jyothibasu et al. are motivated to synthesize a freestanding negative electrode composed of V_2O_5/f-CNT/PPy due to the evidence that conducting polymers including PPy, polyaniline, and PEDOT improve the electrical conductivity of V_2O_5 [101, 102]. By inhibiting the dissolution of V_2O_5, the PPy coating on the V_2O_5/CNT strengthened the cycle life stability and capacitance of the V_2O_5/CNT/PPy composite electrode. With an impressive areal capacitance of $1266\,\text{mF/cm}^2$ and a capacitance retention of 83% after 10,000 charge–discharge cycles, this electrode exhibited exceptional cycling stability, as shown in Fig. 5.17 [103]. Further investigation into improving the electrochemical performance of V_2O_5 is ongoing; nonetheless, it is possible to draw comparisons between the studies that have been reviewed. Graphene significantly improves the electrochemical performance of V_2O_5, as evidenced by the discussion of carbon-based composites including V_2O_5. Superior chemical stability, a substantial specific surface area, and high thermal and electrical conductivity are all characteristics of graphene. It therefore makes an ideal complement to the V_2O_5 electrode material. As a result of the flexible and polymeric nature of conducting polymer and the charge storage and structural and mechanical support that metal oxide provides for conducting polymer, the coexistence of polymer and V_2O_5 results in the manifestation of mutually beneficial features. Particular capacitance was increased by combining polyaniline, one of the most widely used polymers, with V_2O_5.

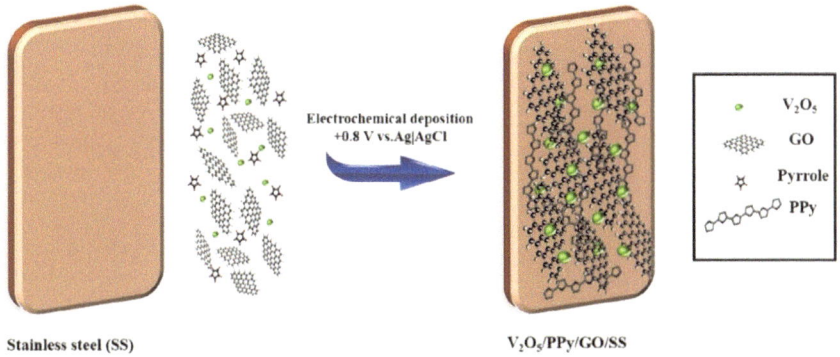

Fig. 5.14 Electrochemical deposition of V_2O_5/PPy/GO nanocomposite [101] (Elsevier Copyright 2017 with license number: 5770540788641)

5.3.6 Vanadium Dioxide (VO₂)

Vanadium dioxide has garnered attention as a potential electrode material for lithium-ion batteries and high-performance supercapacitors on account of its simple fabrication method, low cost, and terrestrial availability [104–106]. Even more significantly, VO_2 possesses a wide potential window and a high capacity for charge storage as a result of its several stable oxidation states [104, 106]. Numerous techniques exist for producing VO_2-based nanocomposites, such as solvothermal synthesis [107, 108], CVD [109], and sputtering [110]. However, similar to other TMO-based electrodes, VO_2-based electrodes have been found to have weak cyclic stability. Efforts are underway to enhance stability through the implementation of composite techniques and nanostructuring to incorporate mechanical strength reinforcements [111]. VO_2 (B), one of the metastable phases of vanadium dioxide, is gaining increasing attention in the field of energy storage technology. The material's layered structure and several oxidation states facilitate charge storage via insertion and enable rapid Faradaic reactions on its surface [112–115]. However, the instability of B-structural VO_2 results in inadequate cyclic stability [111, 116]. In order to mitigate the shortcomings of VO_2, several materials are combined with it, including heteroatoms [117] pore architectures [118], the addition of metal ions [119], the formation of nanocomposites [120], and so on [121]. The polarization effect resulting from the unfavorable shifting of electrostatics along the extended V–V zigzag chain (0.316 nm) in VO_2 monoclinic hinders charge carrying [122–124] in the promising [1×1] tunnel configuration. Potentially resolving these limitations is a monoclinic form of VO_2 that is both stable and devoid of imperfections, possesses a modifiable nanoscale shape, and exhibits encouraging surface activity. Nevertheless, the synthesis of a VO_2-based nanomaterial with structural stability proves to be a formidable task due to the many oxidation states of vanadium [122]. The shape of vanadium dioxide and, consequently, its electrochemical performance was discovered to be influenced by the growing time employed in the solvothermal method.

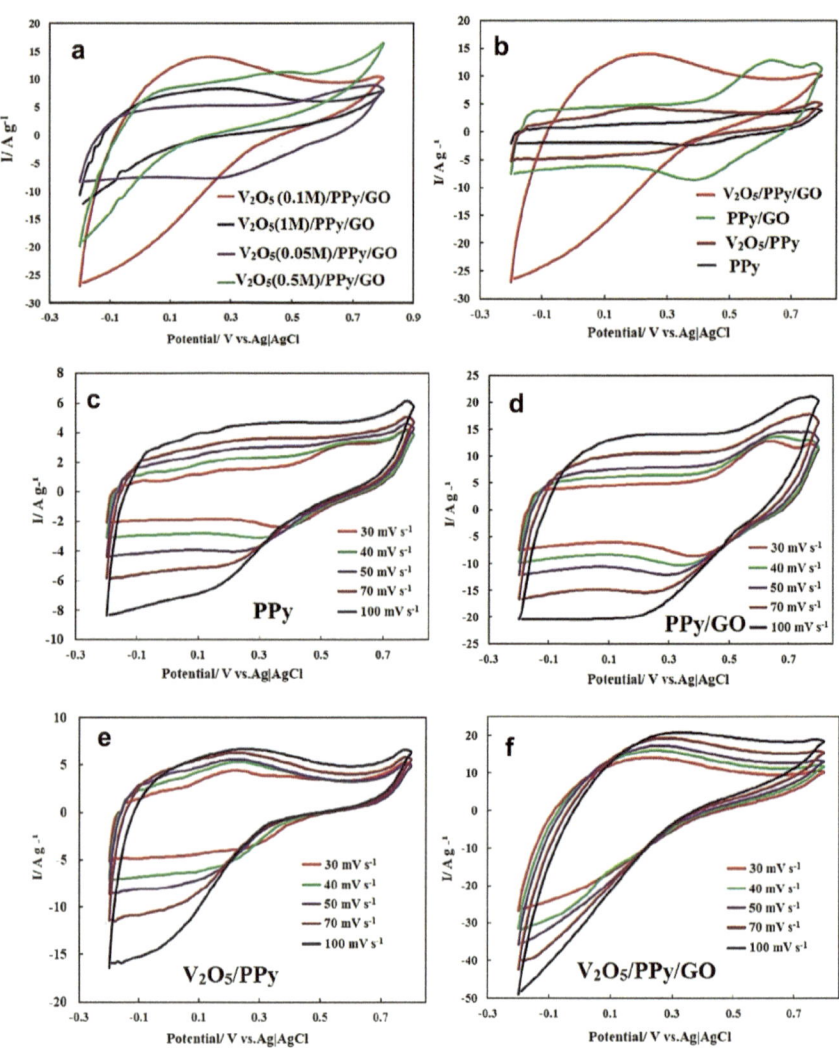

Fig. 5.15 **a** Cyclic voltammograms of V_2O_5/PPy/GO with different vanadyl acetate concentrations and **b** PPy, V_2O_5/PPy, PPy/GO and V_2O_5/PPy/GO in 0.5 M Na_2SO_4 at a scan rate of 30 mV/s, as well as **c** PPy, **d** PPy/GO, **e** V_2O_5/PPy, **f** V_2O_5/PPy/GO at various scan rates in 0.5 M Na_2SO_4 [101] (Elsevier Copyright 2017 with license number: 5770541136746)

The VO_2 samples exhibited a monoclinic crystal structure when prepared for 4 h and 6 h with VO_2 (B) monoclinic phase and 2.5 h with VO_2 (A) monoclinic phase (Fig. 5.18). The morphologies of VO_2 samples grown for 4 h and 6 h, respectively, resembled those of nanoflakes and nanosheets, whereas the morphologies of samples grown for 2.5 h, and 12 h resembled those of nanorods. The six-hour sample was the most effective. By utilizing a current density of 0.5 A.g^{-1} to achieve

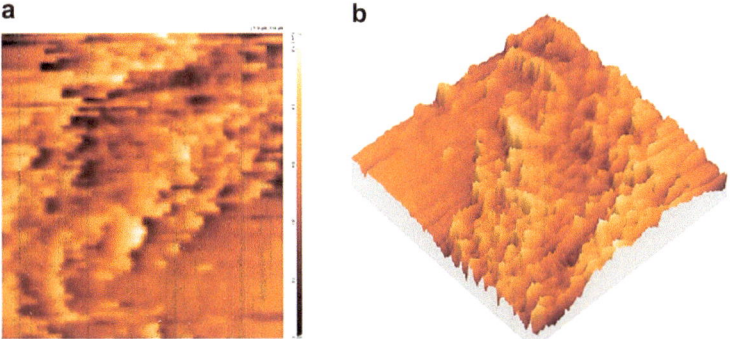

Fig. 5.16 Atomic force microscopy images of **a** 2D and **b** 3D V$_2$O$_5$/PPy/GO/SS electrode [101] (Elsevier Copyright 2017 with license number: 5770550095999)

the highest specific discharge capacity of 49.28 mAh/g and a scan rate of 5 mV/s to measure the corresponding specific capacitance of 663 F.g^{-1}, this specimen exhibited superior electrochemical performance, increased pore volume, and optimal specific surface area (Fig. 5.19). The samples exhibited Columbic efficiency of 99.4% for a maximum of 5000 charge–discharge cycles when subjected to a current density of 10 A.g^{-1} [125]. By employing radio frequency reactive magnetron sputtering at a substrate temperature of 300 °C and manipulating O$_2$ flow rates, I. Neelakanta Reddy et al. successfully deposited monoclinic VO$_2$ nanorod thin films onto glass substrates covered with indium tin oxide. The specific capacitance of the VO$_2$ (M) nanorod thin films was 486 mF/cm at 10 mV/S. In a capacitance test involving 5000 cycles at 100 mV/s, the specific capacitance remained constant at 118 mF/cm [126]. A two-stage hydrothermal process was employed to generate fine 1D VO$_2$ utilizing an innovative synthesis approach. Under the circumstances generated by the hydrothermal reaction, V$_2$O$_5$ was effectively transformed into V$_3$O$_7$, followed by the formation of a fine 1D VO$_2$ monoclinic structure. VO$_2$ exceptional supercapacitive behavior was enabled by its morphological prowess, which enabled it to generate a good binder-free electrode material and display a highly promising surface activity. At 2160 W/kg, the supercapacitance was 15.8 F.g^{-1} and the energy density was 122.3 Wh/kg [122].

Through different treatments of precipitation of precursors, hydrothermal, and calcination, Yifu Zhang et al. created a unique approach with a template-free synthesis of 3D porous VO$_2$ (B) hierarchical solid spheres and hollow spheres. Specific capacitances of 3D VO$_2$ (B) hollow and solid spheres as electrode materials are as high as 1175 mF/cm (336 F.g^{-1}) and 951 mF/cm (272 F.g^{-1}), respectively, when operating at 2 mA/cm. VO$_2$ (B) solid spheres and hollow spheres retained capacitance for approximately 49% and 48%, respectively, after 10,000 cycles. Furthermore, symmetric supercapacitor (SSC) devices were constructed, showcasing commendable pseudocapacitive properties by employing VO$_2$ (B) solid and hollow spheres. The VO$_2$ (B) hollow sphere-SSC device generated an area capacitance of 246 mF/

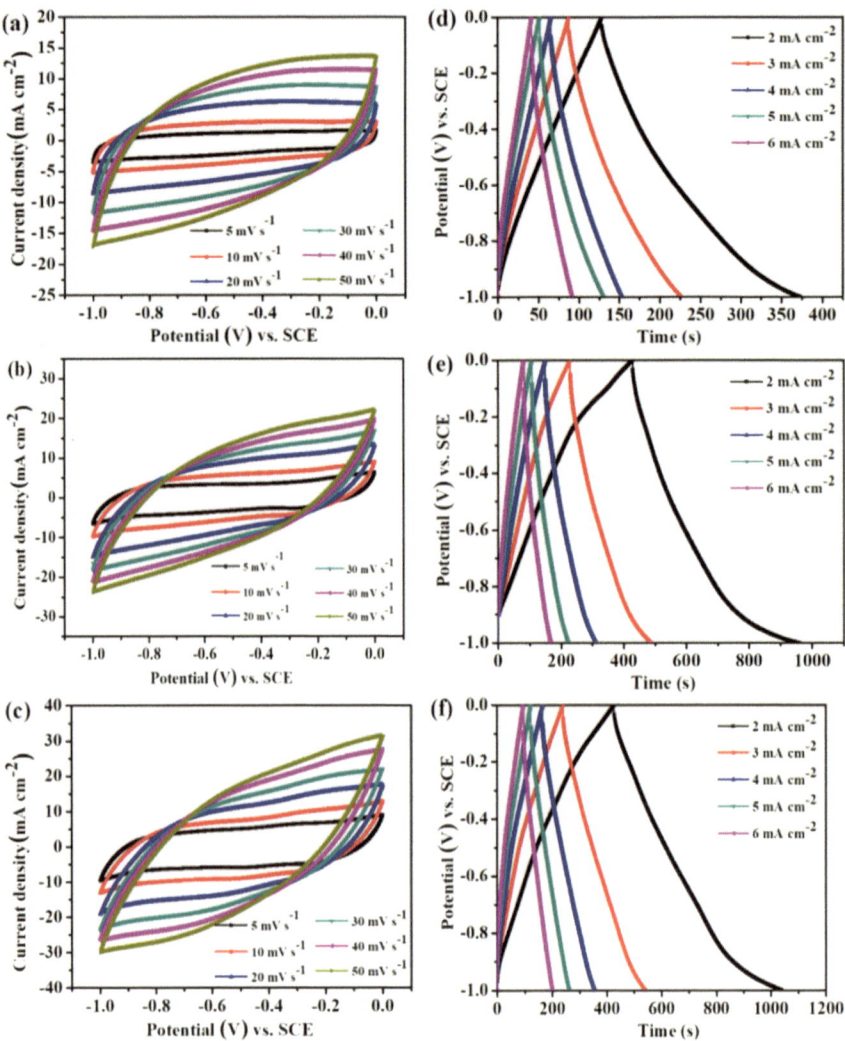

Fig. 5.17 CV curves with different scan rates and GCD curves at different current densities of **a, d** the CNT, **b, e** the V$_2$O$_5$/CNT, and **c, f** the V$_2$O$_5$/CNT/PPy electrodes [103] (Reproduced with permission under CC-BY 4.0)

cm when operated at 1 mA/cm, in contrast to the VO$_2$ (B) solid sphere-SSC device which generated an area capacitance of 122 mF/cm [127].

In addition to the hybridization approach involving conductive materials, which generally lack electrochemical activity, defect engineering has surfaced as a feasible technique for augmenting the redox reaction activities of oxide materials and accelerating electrochemical reaction kinetics [128–130]. As per first-principles calculations, the introduction of a defect into a material not only results in an increase in intrinsic electronic conductivity due to the introduction of additional electrons or

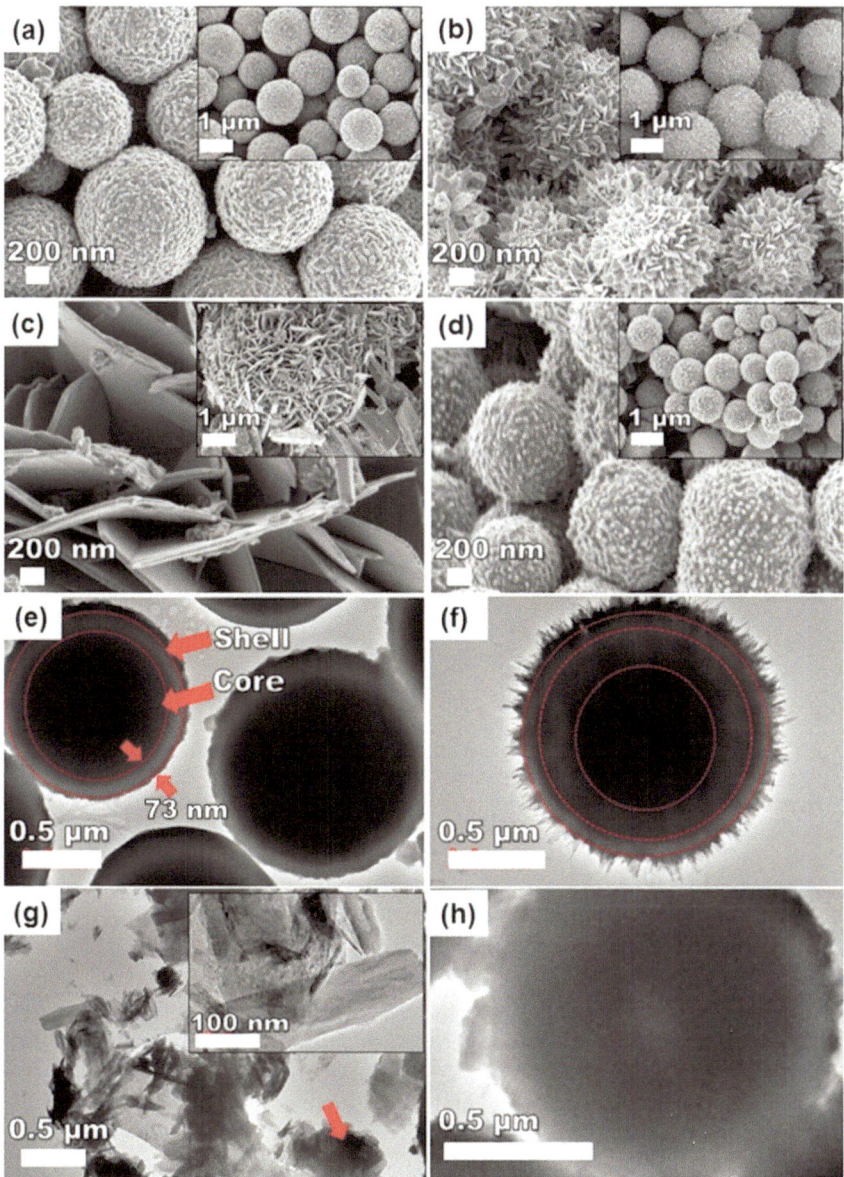

Fig. 5.18 The SEM images of VO$_2$ grown at 200 °C for **a** 2.5 h, **b** 4 h, **c** 6 h, and **d** 12 h with their corresponding TEM images for **e** 2.5 h, **f** 4 h, **g** 6 h, and **h** 12 h [125]. (Elsevier Copyright 2018 with license number: 5771240976481)

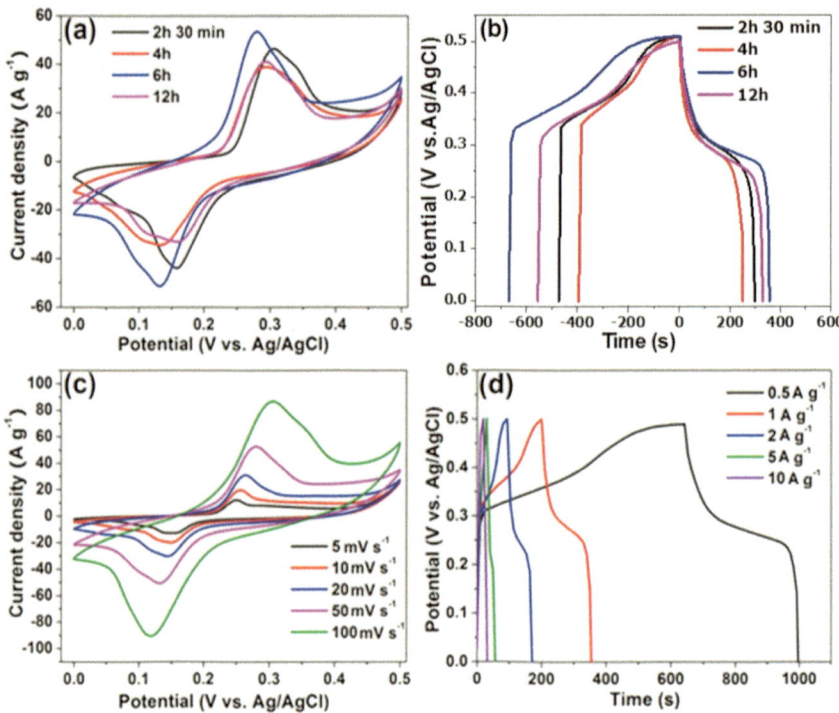

Fig. 5.19 **a** CV curves obtained at a scan rate of 50 mV/s and current density of 0.5 A.g^{-1} as well as their corresponding **b** GCD curves for VO$_2$ grown at 200 °C. **c** CV curves with different scan rates and **d** GCD curves as a function of current density for 6 h-treated VO$_2$ [125]. (Elsevier Copyright 2018 with license number: 5771250060098)

holes, but also diminishes the ion diffusion pathway and the ion transport barrier, thereby accelerating the electrochemical reaction dynamics [128, 131]. Shuai Chen et al. have successfully synthesized monoclinic VO$_2$ nanosheet arrays including modifiable defect domains through the utilization of a straightforward thermal-induced approach. The defect domains are found to be predominantly produced by residual trivalent vanadium originating from the precursor of V$_2$O$_3$, which has a direct influence on the intrinsic conductivity of VO$_2$. Amidst a broad voltage range of -1.5 V to 0.15 V in 15 M LiTFSI, the anode composed of M-d-VO$_2$@CC exhibits a commendable capacitance of 180.3 F.g^{-1} at 0.5 A.g^{-1} and 78.8 F.g^{-1} at 10 A.g^{-1}. High-ordered arrays of titania nanotubes (TNTs) can function as substrates for the deposition of active materials due to their enhanced electrical conductivity, increased surface area, chemical stability, mechanical resilience, and capacity to function without binders. These support structures facilitate the loading of a wide range of meticulously engineered pseudo-active materials via their expansive and perforated tubular channels. As a result, electrode substrates comprised of highly ordered TNT arrays on titanium foil enhance the utilization of active materials by virtue of the increased electrolyte–electrode interface, mitigate the fading effect of electroactive materials, improve cyclic stability, facilitate swift electron-conducting pathways, and offer sturdy mechanical support, among other advantageous attributes [132, 133].

K. M. Thulasi et al. successfully fabricated TNT-VO_2(M) nanocomposites using a series of hydrothermal treatments at different temperatures and potentiostatic anodization, as shown in Fig. 5.20. TNT-V180 exhibited superior electrochemical performance among all three electrode configurations, as evidenced by its specific capacitance values of 152.21 mF/cm and scan rates of 5 mV/s. Furthermore, TNT-V180 exhibited an exceptional Coulombic efficiency and 99% preservation of initial capacitance even after 3000 cycles, validating its suitability for use in supercapacitors [134]. To improve the electrochemical performance of VO_2, it is possible to dope it with another metal. The effective synthesis of Mn-doped VO_2(B) nanosheets was accomplished by Gao et al. using a simple one-step solvothermal method including the direct addition of a manganese salt. Compared to pure VO_2(B) nanosheets, 8.7 mol percent Mn doping results in an 80% increase in specific capacitance, a 2% reduction in charge transfer resistance, and a 2% enhancement in cycle stability. The increased charge storage properties of Mn-doped samples are responsible for the enhanced mass transfer kinetics resulting from stacking with thinner nanosheets and the incorporation of charged defects formed by Mn^{2+} and Mn^{3+} as well as V^{4+} and V^{5+} ions [135]. These defects enhance electrical conductivity and charge transfer efficiency. Furthermore, in the presence of ammonium tungstate, Yifu Zhang et al. employed a straightforward hydrothermal method for producing W-doped VO_2(B) nanobelts via ethanol-reducing peroxovanadium (V) complexes. The VO_2(B) matrix underwent a successful integration of the W atom, leading to the formation of solid solutions consisting entirely of W-doped VO_2(B) nanobelts. The pseudocapacitance qualities of W-doped VO_2(B) nanobelts were exceptional for current densities of 1, 2, 5, 10, and 20 A.g^{-1}. The corresponding specific capacitance values were 253, 239, 207, 164, and 148 F.g^{-1}. Furthermore, they showcased remarkable energy densities of 323, 305, 265, 209, and 189 Wh/kg, in addition to power densities of 2882, 5758, 14,433, 28, 828, and 57,661 W/kg. Various kinds of carbon-based materials have a positive impact on the performance of VO_2. It is acknowledged that carbon fiber felt (CFF) has been a crucial component in supercapacitors as a cost-effective flexible substrate [136] due to its prior modification and activation. The three-dimensional spatial organization of the pores in CFF facilitates efficient energy conversion, rapid ion/electron transport, and the buffering of electrode volume fluctuations caused by metal ion intercalation and deintercalation. By employing CFF, the limitations associated with transition metals in supercapacitors are surmounted [137]. W. J. Zhang et al. modified Carbon fiber felt (CFF) with a three-dimensional structure using chemical bath deposition to manufacture one-dimensional VO_2(B) nanoribbons. The specific capacitance of CFFVO-60 can attain a value of 1248 mF/cm^2 at a current density of 1 mA/cm^2, as shown in Fig. 5.21, a substantial increase compared to the capacitance of the VO_2(B) membrane, which was measured at 451 mF/cm^2. It was observed that 80% of the initial capacitance value is retained after 10^4 cycles of stability [136].

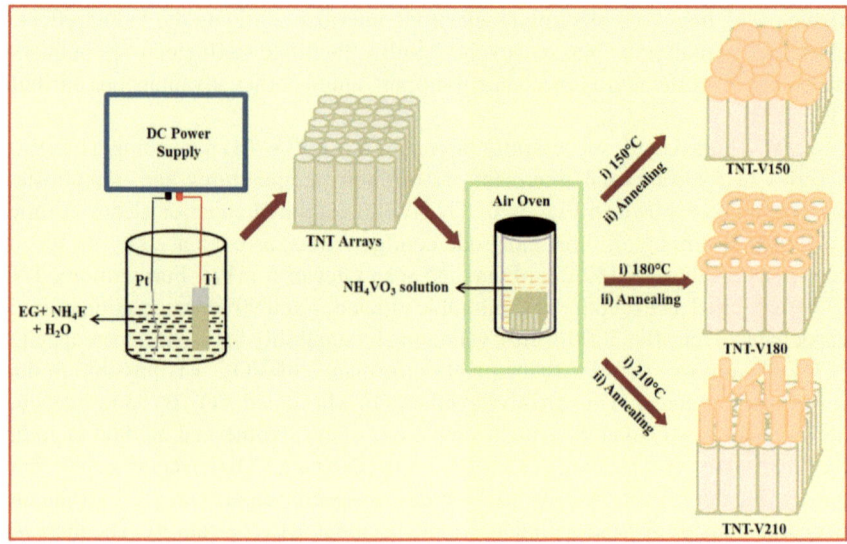

Fig. 5.20 Schematic synthesis steps of TNT-VO$_2$(M) nanocomposites [134] (Elsevier Copyright 2020 with license number: 5771750064658)

5.4   Challenges and Future Perspectives

There has been considerable interest in the advancement of supercapacitors in recent years. Numerous possible developments for the breakthrough technology of super-capacitors have been shown. They possess substantial energy, a very high power density, and a notable capacitance. Determining an optimal electrolyte to enhance the effectiveness of charge storage and attaining a substantial energy density are two challenges that persist in their ongoing advancement. Supercapacitor breakthroughs will be propelled by the research and application of novel electrode materials in conjunction with other substances. Anticipated improvements include increased conductivity, sustainability, redox reaction enrichment, and resolution of challenges such as rapid discharge, limited redox activity, a narrow potential window, and internal resistance have been discussed. Subsequent investigations should be centered on the progression of wearable, self-powered, and lightweight electronics.

Substantially promising potential has been exhibited by materials composed of vanadium oxide, owing to their exceptional stability and high specific capacitance. The following are many crucial factors to contemplate with regard to future perspectives:

1. Investigations into the development of innovative synthesis techniques: Explorations into the implementation of synthesis procedures that utilize electrodeposition, templates, hydrothermal, and sol–gel processes may provide electrode materials exhibiting enhanced morphologies and electrochemical characteristics. However, it is observed that the electrodeposition method is more promising due

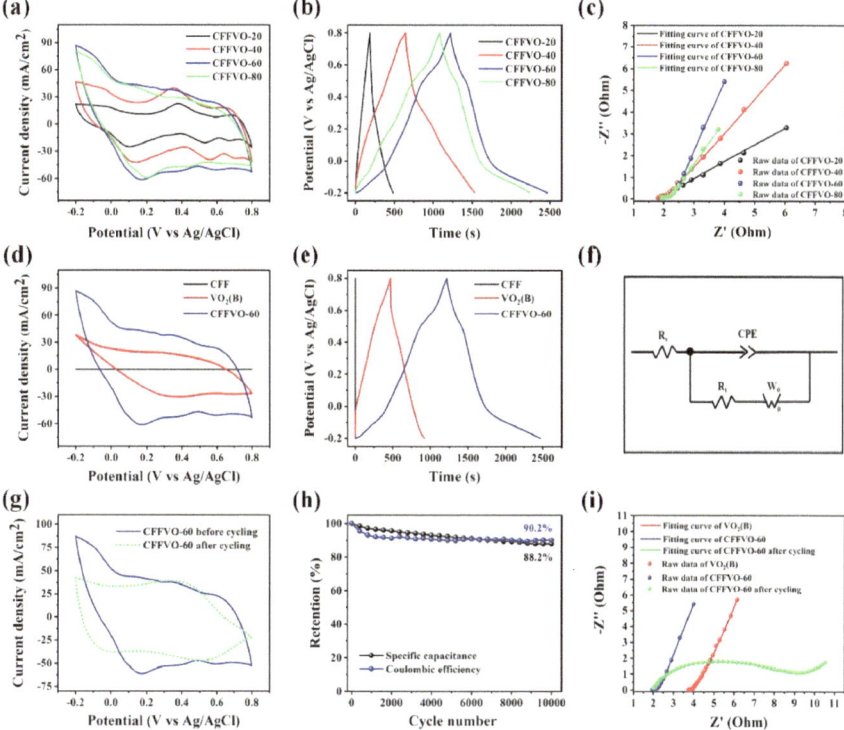

Fig. 5.21 a–c CV, GCD, and EIS spectra of CFFVO-20, 40, 60 and 80 and **d, e** the comparison performances of CV and GCD spectra of pristine CFF, pure VO$_2$(B) and CFFVO-60; **f** the electrical equivalent circuit model of EIS curves in **c** and **i** and electrochemical analysis of CFFVO-60 before and after cycling test: **g** CV curves, **h** specific capacitance, **i** EIS curves [136] (Elsevier Copyright 2022 with license number: 5771760280467)

to the relatively high specific capacitance of its prepared VO$_x$-based electrode material, as compared to other synthesis pathways.

2. The investigation of vanadium oxide in nanostructured configurations, including nanowires, nanoflakes, nanotubes, and nanoparticles, has the potential to enhance ion diffusion velocity and surface area, hence improving energy density and power density.

3. Doping and alloying: The investigation of doping and alloying of vanadium oxide with other elements aims to optimize the overall performance of supercapacitors, increase conductivity, and modify the electrical structure of the material.

4. By combining vanadium oxide with other active substances, such as conducting polymers or carbon-based compounds, it is possible to produce hybrid structures with enhanced energy storage capacities.

5. The investigation of vanadium oxide-based material including flexible and wearable energy storage devices may lead to the development of smart clothing, portable electronics, and wearable electronics.

6. Environmental sustainability: The advancement of supercapacitor development in an environmentally sustainable fashion can be facilitated through research by utilizing precursors or additives in the synthesis of vanadium oxide-based materials that are environmentally friendly and sustainable. In order to effectively scale up and gain widespread use of vanadium oxide-based supercapacitors, more economic and feasibility analyses, process optimization, and adherence to industry standards are necessary.

References

1. W.H. Low, P.S. Khiew, S.S. Lim, C.W. Siong, E.R. Ezeigwe, Recent development of mixed transition metal oxide and graphene/mixed transition metal oxide based hybrid nanostructures for advanced supercapacitors. J. Alloy. Compd. **775**, 1324–1356 (2019)
2. T.S. Kasırga, J.M. Coy, J.H. Park, D.H. Cobden, Visualization of one-dimensional diffusion and spontaneous segregation of hydrogen in single crystals of VO_2. Nanotechnology **27**(34), 345708 (2016)
3. Y. Wang, J. Guo, T. Wang, J. Shao, D. Wang, Y.-W. Yang, Mesoporous transition metal oxides for supercapacitors. Nanomaterials 1667–1689 (2015). [Online]
4. Y. Yan, B. Li, W. Guo, H. Pang, H. Xue, Vanadium based materials as electrode materials for high performance supercapacitors. J. Power. Sources **329**, 148–169 (2016)
5. R.S. Kate, S.A. Khalate, R.J. Deokate, Overview of nanostructured metal oxides and pure nickel oxide (NiO) electrodes for supercapacitors: a review. J. Alloy. Compd. **734**, 89–111 (2018)
6. Y. Zhang, L. Li, H. Su, W. Huang, X. Dong, Binary metal oxide: advanced energy storage materials in supercapacitors. J. Mater. Chem. A **3**(1), 43–59 (2015)
7. R. Liang, Y. Du, P. Xiao, J. Cheng, S. Yuan, Y. Chen, J. Yuan, J. Chen, Transition metal oxide electrode materials for supercapacitors: a review of recent developments. Nanomaterials (2021). [Online]
8. M. Dai, D. Zhao, X. Wu, Research progress on transition metal oxide based electrode materials for asymmetric hybrid capacitors. Chin. Chem. Lett. **31**(9), 2177–2188 (2020)
9. P. Hu, D. Zhao, H. Liu, K. Chen, X. Wu, Engineering PPy decorated $MnCo_2O_4$ urchins for quasi-solid-state hybrid capacitors. CrystEngComm **21**(10), 1600–1606 (2019)
10. Y. Zhang, C. Meng, Facile one-pot hydrothermal synthesis of belt-like β-V_6O_{13} with rectangular cross sections for Li-ion battery application. Mater. Lett. **160**, 404–407 (2015)
11. M.S. Shalaby, M.O. Alziyadi, H. Gamal, S. Hamdy, Solid-state lithium-ion battery: the key components enhance the performance and efficiency of anode, cathode, and solid electrolytes. J. Alloy. Compd. **969**, 172318 (2023)
12. M. Lee, S.K. Balasingam, H.Y. Jeong, W.G. Hong, H.-B.-R. Lee, B.H. Kim, Y. Jun, One-step hydrothermal synthesis of graphene decorated V_2O_5 nanobelts for enhanced electrochemical energy storage. Sci. Rep. **5**(1), 8151 (2015)
13. Y. Zhang, N. Wang, Y. Huang, W. Wu, C. Huang, C. Meng, Fabrication and catalytic activity of ultra-long V_2O_5 nanowires on the thermal decomposition of ammonium perchlorate. Ceram. Int. **40**(7, Part B), 11393–11398 (2014)
14. W. Yan, M. Hu, D. Wang, C. Li, Room temperature gas sensing properties of porous silicon/ V_2O_5 nanorods composite. Appl. Surf. Sci. **346**, 216–222 (2015)
15. A. Roy, A. Ray, P. Sadhukhan, S. Saha, S. Das, Morphological behaviour, electronic bond formation and electrochemical performance study of V_2O_5-polyaniline composite and its application in asymmetric supercapacitor. Mater. Res. Bull. **107**, 379–390 (2018)

16. K. Panigrahi, P. Howli, K.K. Chattopadhyay, 3D network of V_2O_5 for flexible symmetric supercapacitor. Electrochim. Acta **337**, 135701 (2020)
17. D.J. Ahirrao, K. Mohanapriya, N. Jha, V_2O_5 nanowires-graphene composite as an outstanding electrode material for high electrochemical performance and long-cycle-life supercapacitor. Mater. Res. Bull. **108**, 73–82 (2018)
18. V.P. Prasadam, N. Bahlawane, F. Mattelaer, G. Rampelberg, C. Detavernier, L. Fang, Y. Jiang, K. Martens, I.P. Parkin, I. Papakonstantinou, Atomic layer deposition of vanadium oxides: process and application review. Mater. Today Chem. **12**, 396–423 (2019)
19. P. Hu, P. Hu, T.D. Vu, M. Li, S. Wang, Y. Ke, X. Zeng, L. Mai, Y. Long, Vanadium oxide: phase diagrams, structures, synthesis, and applications. Chem. Rev. **123**(8), 4353–4415 (2023)
20. E. Baudrin, G. Sudant, D. Larcher, B. Dunn, J.-M. Tarascon, Preparation of nanotextured $VO_2[B]$ from vanadium oxide aerogels. Chem. Mater. **19**(8), 2140–2140 (2007)
21. T.M. Benedetti, E. Redston, W.G. Menezes, D.M. Reis, J.F. Soares, A.J.G. Zarbin, R.M. Torresi, Lithium intercalation in nanostructured thin films of a mixed-valence layered vanadium oxide using an ionic liquid electrolyte. J. Power. Sources **224**, 72–79 (2013)
22. M.-L. Fu, G.-C. Guo, A.-Q. Wu, B. Liu, L.-Z. Cai, J.-S. Huang, Synthesis, crystal structure and characterisation of a novel chiral mixed-valence vanadium oxide hybrid, $[V_5O_{11}(dicn)_3]$. Eur. J. Inorg. Chem. **2005**(15), 3104–3108 (2005)
23. F. Lanlan, L. Zhenhuan, D. Nanping, Recent advances in vanadium-based materials for aqueous metal ion batteries: design of morphology and crystal structure, evolution of mechanisms and electrochemical performance. Energy Storage Mater. **41**, 152–182 (2021)
24. C. Jing, X.D. Liu, K. Li, X. Liu, B. Dong, F. Dong, Y. Zhang, The pseudocapacitance mechanism of graphene/CoAl LDH and its derivatives: are all the modifications beneficial? J. Energy Chem. **52**, 218–227 (2021)
25. C. Jing, B. Dong, Y. Zhang, Chemical modifications of layered double hydroxides in the supercapacitor. Energy Environ. Mater. **3**(3), 346–379 (2020)
26. X. Li, D. Du, Y. Zhang, W. Xing, Q. Xue, Z. Yan, Layered double hydroxides toward high-performance supercapacitors. J. Mater. Chem. A **5**(30), 15460–15485 (2017)
27. C. Jing, X. Song, K. Li, Y. Zhang, X. Liu, B. Dong, F. Dong, S. Zhao, H. Yao, Y. Zhang, Optimizing the rate capability of nickel cobalt phosphide nanowires on graphene oxide by the outer/inter-component synergistic effects. J. Mater. Chem. A **8**(4), 1697–1708 (2020)
28. D. Chen, X. Rui, Q. Zhang, H. Geng, L. Gan, W. Zhang, C. Li, S. Huang, Y. Yu, Persistent zinc-ion storage in mass-produced V_2O_5 architectures. Nano Energy **60**, 171–178 (2019)
29. S. Zhang, H. Tan, X. Rui, Y. Yu, Vanadium-based materials: next generation electrodes powering the battery revolution? Acc. Chem. Res. **53**(8), 1660–1671 (2020)
30. Y. Li, M. Chen, B. Liu, Y. Zhang, X. Liang, X. Xia, Heteroatom doping: an effective way to boost sodium ion storage. Adv. Energy Mater. **10**(27), 2000927 (2020)
31. C. Liu, R. Massé, X. Nan, G. Cao, A promising cathode for Li-ion batteries: $Li_3V_2(PO_4)_3$. Energy Storage Mater. **4**, 15–58 (2016)
32. R.C. Massé, C. Liu, Y. Li, L. Mai, G. Cao, Energy storage through intercalation reactions: electrodes for rechargeable batteries. Natl. Sci. Rev. **4**(1), 26–53 (2016)
33. R.K. Guduru, J.C. Icaza, A brief review on multivalent intercalation batteries with aqueous electrolytes. Nanomaterials (2016). [Online]
34. Z. Wang, M. Zhang, W. Ma, J. Zhu, W. Song, Application of carbon materials in aqueous zinc ion energy storage devices. Small **17**(19), 2100219 (2021)
35. R. Li, F. Xing, T. Li, H. Zhang, J. Yan, Q. Zheng, X. Li, Intercalated polyaniline in V_2O_5 as a unique vanadium oxide bronze cathode for highly stable aqueous zinc ion battery. Energy Storage Mater. **38**, 590–598 (2021)
36. S. Chen, K. Li, K.S. Hui, J. Zhang, Regulation of lamellar structure of vanadium oxide via polyaniline intercalation for high-performance aqueous zinc-ion battery. Adv. Func. Mater. **30**(43), 2003890 (2020)
37. W. Li, C. Han, Q. Gu, S.-L. Chou, J.-Z. Wang, H.-K. Liu, S.-X. Dou, Electron delocalization and dissolution-restraint in vanadium oxide superlattices to boost electrochemical performance of aqueous zinc-ion batteries. Adv. Energy Mater. **10**(48), 2001852 (2020)

38. X.H. Rui, C. Li, J. Liu, T. Cheng, C.H. Chen, The $Li_3V_2(PO_4)$ 3/C composites with high-rate capability prepared by a maltose-based sol–gel route. Electrochim. Acta **55**(22), 6761–6767 (2010)

39. D. Kong, X. Li, Y. Zhang, X. Hai, B. Wang, X. Qiu, Q. Song, Q.-H. Yang, L. Zhi, Encapsulating V_2O_5 into carbon nanotubes enables the synthesis of flexible high-performance lithium ion batteries. Energy Environ. Sci. **9**(3), 906–911 (2016)

40. W. Ren, Z. Zheng, C. Xu, C. Niu, Q. Wei, Q. An, K. Zhao, M. Yan, M. Qin, L. Mai, Self-sacrificed synthesis of three-dimensional $Na_3V_2(PO_4)_3$ nanofiber network for high-rate sodium–ion full batteries. Nano Energy **25**, 145–153 (2016)

41. Y. Jiang, X. Zhou, D. Li, X. Cheng, F. Liu, Y. Yu, Highly reversible Na storage in $Na_3V_2(PO_4)_3$ by optimizing nanostructure and rational surface engineering. Adv. Energy Mater. **8**(16), 1800068 (2018)

42. H. Zhang, X. Han, R. Gan, Z. Guo, Y. Ni, L. Zhang, A facile biotemplate-assisted synthesis of mesoporous V_2O_5 microtubules for high performance asymmetric supercapacitors. Appl. Surf. Sci. **511**, 145527 (2020)

43. X. Yang, L. Zhang, F. Zhang, T. Zhang, Y. Huang, Y. Chen, A high-performance all-solid-state supercapacitor with graphene-doped carbon material electrodes and a graphene oxide-doped ion gel electrolyte. Carbon **72**, 381–386 (2014)

44. O. Koysuren, C. Du, N. Pan, G. Bayram, Preparation and comparison of two electrodes for supercapacitors: Pani/CNT/Ni and Pani/Alizarin-treated nickel. J. Appl. Polym. Sci. **113**(2), 1070–1081 (2009)

45. V. Khomenko, E. Frackowiak, F. Béguin, Determination of the specific capacitance of conducting polymer/nanotubes composite electrodes using different cell configurations. Electrochim. Acta **50**(12), 2499–2506 (2005)

46. S. Saha, P. Samanta, N.C. Murmu, T. Kuila, A review on the heterostructure nanomaterials for supercapacitor application. J. Energy Storage **17**, 181–202 (2018)

47. Z. Yu, L. Tetard, L. Zhai, J. Thomas, Supercapacitor electrode materials: nanostructures from 0 to 3 dimensions. Energy Environ. Sci. **8**(3), 702–730 (2015)

48. B. Saravanakumar, K.K. Purushothaman, G. Muralidharan, High performance supercapacitor based on carbon coated V_2O_5 nanorods. J. Electroanal. Chem. **758**, 111–116 (2015)

49. Y. Zhang, J. Zheng, Q. Wang, T. Hu, F. Tian, C. Meng, Facile preparation, optical and electrochemical properties of layer-by-layer V_2O_5 quadrate structures. Appl. Surf. Sci. **399**, 151–159 (2017)

50. Y. Guo, J. Li, M. Chen, G. Gao, Facile synthesis of vanadium pentoxide@carbon core–shell nanowires for high-performance supercapacitors. J. Power. Sources **273**, 804–809 (2015)

51. Y. Zhang, M. Fan, X. Liu, C. Huang, H. Li, Beltlike V_2O_3@C core–shell-structured composite: design, preparation, characterization, phase transition, and improvement of electrochemical properties of V_2O_3. Eur. J. Inorg. Chem. **2012**(10), 1650–1659 (2012)

52. J. Zheng, T. Hu, Y. Zhang, T. Lv, F. Tian, C. Meng, Template-free synthesis of porous V_2O_5 flakes as a battery-type electrode material with high capacity for supercapacitors. Colloids Surf. A **553**, 317–326 (2018)

53. G. Silversmit, D. Depla, H. Poelman, G.B. Marin, R. De Gryse, Determination of the V2p XPS binding energies for different vanadium oxidation states (V^{5+} to V^{0+}). J. Electron Spectrosc. Relat. Phenom. **135**(2), 167–175 (2004)

54. J. Yang, T. Lan, J. Liu, Y. Song, M. Wei, Supercapacitor electrode of hollow spherical V_2O_5 with a high pseudocapacitance in aqueous solution. Electrochim. Acta **105**, 489–495 (2013)

55. Y. Zhang, C. Huang, C. Meng, T. Hu, A novel route for synthesis and growth formation of metal oxides microspheres: Insights from V_2O_3 microspheres. Mater. Chem. Phys. **177**, 543–553 (2016)

56. M. Li, T. Ai, L. Kou, J. Song, W. Bao, Y. Wang, X. Wei, W. Li, Z. Deng, X. Zou, H. Wang, Synthesis and electrochemical performance of V_2O_5 nanosheets for supercapacitor. AIP Adv. **12**(5), 055203 (2022)

57. C. Peng, M. Jin, D. Han, X. Liu, L. Lai, Structural engineering of V_2O_5 nanobelts for flexible supercapacitors. Mater. Lett. **320**, 132391 (2022)

58. B. Balamuralitharan, I.-H. Cho, J.-S. Bak, H.-J. Kim, V_2O_5 nanorod electrode material for enhanced electrochemical properties by a facile hydrothermal method for supercapacitor applications. New J. Chem. **42**(14), 11862–11868 (2018)
59. S.S. Karade, S. Lalwani, J.-H. Eum, H. Kim, Coin cell fabricated symmetric supercapacitor device of two-steps synthesized V_2O_5 Nanorods. J. Electroanal. Chem. **864**, 114080 (2020)
60. Y. Zhang, J. Zheng, Y. Zhao, T. Hu, Z. Gao, C. Meng, Fabrication of V_2O_5 with various morphologies for high-performance electrochemical capacitor. Appl. Surf. Sci. **377**, 385–393 (2016)
61. D.Y. Kim, S.H. Ju, H.Y. Koo, S.K. Hong, Y.C. Kang, Synthesis of nanosized Co_3O_4 particles by spray pyrolysis. J. Alloy. Compd. **417**(1), 254–258 (2006)
62. Z. Yin, J. Xu, Y. Ge, Q. Jiang, Y. Zhang, Y. Yang, Y. Sun, S. Hou, Y. Shang, Y. Zhang, Synthesis of V_2O_5 microspheres by spray pyrolysis as cathode material for supercapacitors. Mater. Res. Express **5**(3), 036306 (2018)
63. Y. Zhang, Synthesis and characterization of hollow V_2O_5 microspheres for supercapacitor electrode with pseudocapacitance. Mater. Sci.-Pol. **35**, 188–196 (2017)
64. H. Gamal, A.M. Elshahawy, S.S. Medany, M.A. Hefnawy, M.S. Shalaby, Recent advances of vanadium oxides and their derivatives in supercapacitor applications: a comprehensive review. J. Energy Storage **76**, 109788 (2024)
65. J. Mu, J. Wang, J. Hao, P. Cao, S. Zhao, W. Zeng, B. Miao, S. Xu, Hydrothermal synthesis and electrochemical properties of V_2O_5 nanomaterials with different dimensions. Ceram. Int. **41**(10, Part A), 12626–12632 (2015)
66. Y. Zhang, J. Zheng, T. Hu, F. Tian, C. Meng, Synthesis and supercapacitor electrode of VO_2(B)/C core–shell composites with a pseudocapacitance in aqueous solution. Appl. Surf. Sci. **371**, 189–195 (2016)
67. M.D. Patil, S.D. Dhas, A.A. Mane, A.V. Moholkar, Clinker-like V_2O_5 nanostructures anchored on 3D Ni-foam for supercapacitor application. Mater. Sci. Semicond. Process. **133**, 105978 (2021)
68. M.S. Javed, T. Najim, I. Hussain, S. Batool, M. Idrees, A. Mehmood, M. Imran, M.A. Assiri, A. Ahmad, S.S. Ahmad Shah, 2D V_2O_5 nanoflakes as a binder-free electrode material for high-performance pseudocapacitor. Ceram. Int. **47**(17), 25152–25157 (2021)
69. L. Hua, Z. Ma, P. Shi, L. Li, K. Rui, J. Zhou, X. Huang, X. Liu, J. Zhu, G. Sun, W. Huang, Ultrathin and large-sized vanadium oxide nanosheets mildly prepared at room temperature for high performance fiber-based supercapacitors. J. Mater. Chem. A **5**(6), 2483–2487 (2017)
70. T. Peng, T. Zhao, Q. Zhou, H. Zhou, J. Wang, J. Liu, Q. Liu, Design of mass-controllable $NiCo_2S_4$/Ketjen Black nanocomposite electrodes for high performance supercapacitors. CrystEngComm **17**(39), 7583–7591 (2015)
71. C. Zhong, Y. Deng, W. Hu, J. Qiao, L. Zhang, J. Zhang, A review of electrolyte materials and compositions for electrochemical supercapacitors. Chem. Soc. Rev. **44**(21), 7484–7539 (2015)
72. M. Jayachandran, A. Rose, T. Maiyalagan, N. Poongodi, T. Vijayakumar, Effect of various aqueous electrolytes on the electrochemical performance of V_2O_5 spindle-like nanostructures as electrode material for supercapacitor application. J. Mater. Sci. Mater. Electron. **32**(5), 6623–6635 (2021)
73. B. Akkinepally, I. Neelakanta Reddy, C. Lee, T. Jo Ko, P. Srinivasa Rao, J. Shim, Promising electrode material of Fe_3O_4 nanoparticles decorated on V_2O_5 nanobelts for high-performance symmetric supercapacitors. Ceram. Int. **49**(4), 6280–6288 (2023)
74. A. Ray, A. Roy, P. Sadhukhan, S.R. Chowdhury, P. Maji, S.K. Bhattachrya, S. Das, Electrochemical properties of TiO_2-V_2O_5 nanocomposites as a high performance supercapacitors electrode material. Appl. Surf. Sci. **443**, 581–591 (2018)
75. X. Xie, W. Liu, L. Zhao, C. Huang, Structural and electrochemical behavior of Mn–V oxide synthesized by a novel precipitation method. J. Solid State Electrochem. **14**(9), 1585–1594 (2010)
76. A. Lourenco, E. Masetti, F. Decker, Electrochemical and optical characterization of RF-sputtered thin films of vanadium–nickel mixed oxides. Electrochim. Acta **46**(13), 2257–2262 (2001)

77. K. Jeyalakshmi, S. Vijayakumar, K.K. Purushothaman, G. Muralidharan, Nanostructured nickel doped β-V_2O_5 thin films for supercapacitor applications. Mater. Res. Bull. **48**(7), 2578–2582 (2013)
78. F. Coustier, G. Jarero, S. Passerini, W.H. Smyrl, Performance of copper-doped V_2O_5 xerogel in coin cell assembly. J. Power. Sources **83**(1), 9–14 (1999)
79. F. Zhang, J. Ma, H. Yao, Ultrathin Ni–MOF nanosheet coated $NiCo_2O_4$ nanowire arrays as a high-performance binder-free electrode for flexible hybrid supercapacitors. Ceram. Int. **45**(18, Part A), 24279–24287 (2019)
80. N. Guru Prakash, M. Dhananjaya, B. Purusottam Reddy, K. Sivajee Ganesh, A. Lakshmi Narayana, O.M. Hussain, Molybdenum doped V_2O_5 thin films electrodes for supercapacitors. Mater. Today: Proc. 3(10, Part B), 4076–4081 (2016)
81. V. Uma Shankar, D. Govindarajan, P. Christuraj, M.J. Salethraj, F.J. Johanson, M.D. Raja, Enhanced the electrochemical properties of Ni doped V_2O_5 as a electrode material for supercapacitor applications. Mater. Today: Proc. **50**, 2675–2678 (2022)
82. M. Kandasamy, S. Sahoo, S.K. Nayak, B. Chakraborty, C.S. Rout, Recent advances in engineered metal oxide nanostructures for supercapacitor applications: experimental and theoretical aspects. J. Mater. Chem. A **9**(33), 17643–17700 (2021)
83. J.W. Lee, S.Y. Lim, H.M. Jeong, T.H. Hwang, J.K. Kang, J.W. Choi, Extremely stable cycling of ultra-thin V_2O_5 nanowire–graphene electrodes for lithium rechargeable battery cathodes. Energy Environ. Sci. **5**(12), 9889–9894 (2012)
84. S.D. Perera, B. Patel, N. Nijem, K. Roodenko, O. Seitz, J.P. Ferraris, Y.J. Chabal, K.J. Balkus Jr., Vanadium oxide nanowire-carbon nanotube binder-free flexible electrodes for supercapacitors. Adv. Energy Mater. **1**(5), 936–945 (2011)
85. W. Sun, W. Gao, Y. Du, K. Zhang, G. Wu, A facile strategy for fabricating hierarchical nanocomposites of V_2O_5 nanowire arrays on a three-dimensional N-doped graphene aerogel with a synergistic effect for supercapacitors. J. Mater. Chem. A **6**(21), 9938–9947 (2018)
86. K. Wang, H. Wu, Y. Meng, Z. Wei, Conducting polymer nanowire arrays for high performance supercapacitors. Small **10**(1), 14–31 (2014)
87. M. Rasoulis, D. Vernardou, Electrodeposition of vanadium oxides at room temperature as cathodes in lithium-ion batteries. Coatings **7**(7), 100 (2017)
88. H. An, J. Mike, K.A. Smith, L. Swank, Y.-H. Lin, S.L. Pesek, R. Verduzco, J.L. Lutkenhaus, Highly flexible self-assembled V_2O_5 cathodes enabled by conducting diblock copolymers. Sci. Rep. **5**(1), 14166 (2015)
89. N. Kumar, S.-B. Kim, S.-Y. Lee, S.-J. Park, Recent advanced supercapacitor: a review of storage mechanisms, electrode materials, modification, and perspectives. Nanomaterials **12**(20), 3708 (2022)
90. T. Chen, L. Dai, Carbon nanomaterials for high-performance supercapacitors. Mater. Today **16**(7), 272–280 (2013)
91. A. González, E. Goikolea, J.A. Barrena, R. Mysyk, Review on supercapacitors: technologies and materials. Renew. Sustain. Energy Rev. **58**, 1189–1206 (2016)
92. D. Majumdar, M. Mandal, S.K. Bhattacharya, V_2O_5 and its carbon-based nanocomposites for supercapacitor applications. ChemElectroChem **6**(6), 1623–1648 (2019)
93. Q. Ke, J. Wang, Graphene-based materials for supercapacitor electrodes—a review. J. Materiomics **2**(1), 37–54 (2016)
94. E. Aawani, N. Memarian, H.R. Dizaji, Synthesis and characterization of reduced graphene oxide–V_2O_5 nanocomposite for enhanced photocatalytic activity under different types of irradiation. J. Phys. Chem. Solids **125**, 8–15 (2019)
95. S. Stankovich, D.A. Dikin, G.H.B. Dommett, K.M. Kohlhaas, E.J. Zimney, E.A. Stach, R.D. Piner, S.T. Nguyen, R.S. Ruoff, Graphene-based composite materials. Nature **442**(7100), 282–286 (2006)
96. M. Fu, Q. Zhuang, Z. Zhu, Z. Zhang, W. Chen, Q. Liu, H. Yu, Facile synthesis of V_2O_5/graphene composites as advanced electrode materials in supercapacitors. J. Alloy. Compd. **862**, 158006 (2021)

97. A. Aamir, A. Ahmad, S.K. Shah, N.U. Ain, M. Mehmood, Y. Khan, Z.U. Rehman, Electro-codeposition of V_2O_5-polyaniline composite on Ni foam as an electrode for supercapacitor. J. Mater. Sci. Mater. Electron. **31**(23), 21035–21045 (2020)

98. S.-Y. Lee, J.-I. Kim, S.-J. Park, Activated carbon nanotubes/polyaniline composites as supercapacitor electrodes. Energy **78**, 298–303 (2014)

99. B.K. Kuila, B. Nandan, M. Böhme, A. Janke, M. Stamm, Vertically oriented arrays of polyaniline nanorods and their super electrochemical properties. Chem. Commun. **38**, 5749–5751 (2009)

100. M.-H. Bai, T.-Y. Liu, F. Luan, Y. Li, X.-X. Liu, Electrodeposition of vanadium oxide–polyaniline composite nanowire electrodes for high energy density supercapacitors. J. Mater. Chem. A **2**(28), 10882–10888 (2014)

101. P. Asen, S. Shahrokhian, A. Iraji zad, One step electrodeposition of V_2O_5/polypyrrole/graphene oxide ternary nanocomposite for preparation of a high performance supercapacitor. Int. J. Hydrogen Energy **42**(33), 21073–21085 (2017)

102. C.X. Guo, G. Yilmaz, S. Chen, S. Chen, X. Lu, Hierarchical nanocomposite composed of layered V_2O_5/PEDOT/MnO_2 nanosheets for high-performance asymmetric supercapacitors. Nano Energy **12**, 76–87 (2015)

103. J.P. Jyothibasu, M.-Z. Chen, Y.-C. Tien, C.-C. Kuo, E.-C. Chen, Y.-C. Lin, T.-C. Chiang, R.-H. Lee, V_2O_5/carbon nanotube/polypyrrole based freestanding negative electrodes for high-performance supercapacitors. Catalysts **11**(8), 980 (2021)

104. P. Man, Q. Zhang, J. Sun, J. Guo, X. Wang, Z. Zhou, B. He, Q. Li, L. Xie, J. Zhao, C. Li, Q. Li, Y. Yao, Hierarchically structured VO_2@PPy core-shell nanowire arrays grown on carbon nanotube fibers as advanced cathodes for high-performance wearable asymmetric supercapacitors. Carbon **139**, 21–28 (2018)

105. S. Fleischmann, M. Zeiger, N. Jäckel, B. Krüner, V. Lemkova, M. Widmaier, V. Presser, Tuning pseudocapacitive and battery-like lithium intercalation in vanadium dioxide/carbon onion hybrids for asymmetric supercapacitor anodes. J. Mater. Chem. A **5**(25), 13039–13051 (2017)

106. S. Li, G. Liu, J. Liu, Y. Lu, Q. Yang, L.-Y. Yang, H.-R. Yang, S. Liu, M. Lei, M. Han, Carbon fiber cloth@VO_2 (B): excellent binder-free flexible electrodes with ultrahigh mass-loading. J. Mater. Chem. A **4**(17), 6426–6432 (2016)

107. N.B. Mahadi, J.-S. Park, J.-H. Park, K.Y. Chung, S.Y. Yi, Y.-K. Sun, S.-T. Myung, Vanadium dioxide—reduced graphene oxide composite as cathode materials for rechargeable Li and Na batteries. J. Power. Sources **326**, 522–532 (2016)

108. A.A. Boochakravarthy, M. Dhanasekar, S.V. Bhat, Thermochromic behavior of VO_2/GO and VO_2/rGO nanocomposites prepared by a facile hydrothermal method. AIP Advan. 10(8) (2020)

109. H. Kim, Y. Kim, T. Kim, A.R. Jang, H.Y. Jeong, S.H. Han, D.H. Yoon, H.S. Shin, D.J. Bae, K.S. Kim, W.S. Yang, Enhanced optical response of hybridized VO_2/graphene films. Nanoscale **5**(7), 2632–2636 (2013)

110. H. Zhou, J. Li, Y. Xin, X. Cao, S. Bao, P. Jin, Electron transfer induced thermochromism in a VO_2–graphene–Ge heterostructure. J. Mater. Chem. C **3**(19), 5089–5097 (2015)

111. G. Ren, R. Zhang, Z. Fan, VO_2 nanoparticles on edge oriented graphene foam for high rate lithium ion batteries and supercapacitors. Appl. Surf. Sci. **441**, 466–473 (2018)

112. W. Lv, C. Yang, G. Meng, R. Zhao, A. Han, R. Wang, J. Liu, VO_2(B) nanobelts/reduced graphene oxide composites for high-performance flexible all-solid-state supercapacitors. Sci. Rep. **9**(1), 10831 (2019)

113. H.-Y. Li, C. Wei, L. Wang, Q.-S. Zuo, X. Li, B. Xie, Hierarchical vanadium oxide microspheres forming from hyperbranched nanoribbons as remarkably high performance electrode materials for supercapacitors. J. Mater. Chem. A **3**(45), 22892–22901 (2015)

114. C.N. Schmidt, G. Cao, Properties of mesoporous carbon modified carbon felt for anode of all-vanadium redox flow battery. Sci. China Mater. **59**(12), 1037–1050 (2016)

115. Z. Khan, P. Singh, S.A. Ansari, S.R. Manippady, A. Jaiswal, M. Saxena, VO_2 Nanostructures for batteries and supercapacitors: a review. Small **17**(4), 2006651 (2021)

116. A. Moretti, S. Passerini, Bilayered nanostructured $V_2O_5 \cdot nH_2O$ for metal batteries. Adv. Energy Mater. **6**(23), 1600868 (2016)
117. S. Ghosh, S. Barg, S.M. Jeong, K. Ostrikov, Heteroatom-doped and oxygen-functionalized nanocarbons for high-performance supercapacitors. Adv. Energy Mater. **10**(32), 2001239 (2020)
118. Y. Yan, G. Chen, P. She, G. Zhong, W. Yan, B.Y. Guan, Y. Yamauchi, Mesoporous nanoarchitectures for electrochemical energy conversion and storage. Adv. Mater. **32**(44), 2004654 (2020)
119. N. Wang, C. Sun, X. Liao, Y. Yuan, H. Cheng, Q. Sun, B. Wang, X. Pan, K. Zhao, Q. Xu, X. Lu, J. Lu, Reversible (De)intercalation of hydrated Zn^{2+} in Mg^{2+}-stabilized V_2O_5 nanobelts with high areal capacity. Adv. Energy Mater. **10**(41), 2002293 (2020)
120. J. Zhao, H. Li, C. Li, Q. Zhang, J. Sun, X. Wang, J. Guo, L. Xie, J. Xie, B. He, Z. Zhou, C. Lu, W. Lu, G. Zhu, Y. Yao, MOF for template-directed growth of well-oriented nanowire hybrid arrays on carbon nanotube fibers for wearable electronics integrated with triboelectric nanogenerators. Nano Energy **45**, 420–431 (2018)
121. Q. He, Z. Peng, S. Li, L. Tan, Y. Chen, High-energy aqueous asymmetric supercapacitors via synergistic design of electrodes derived from hierarchical vanadium dioxide nanocomposites. ChemElectroChem **9**(2), e202101576 (2022)
122. N. Kumar, A. Juyal, V. Gajraj, S. Upadhyay, N. Priyadarshi, S. Chetana, N. Chandra Joshi, A. Sen, Facile synthesis of fine 1D VO_2 and its supercapacitance as a binder free electrode. Inorg. Chem. Commun. **150**, 110439 (2023)
123. Q. Wang, Y. Zhang, J. Xiao, H. Jiang, X. Li, C. Meng, A novel ordered hollow spherical nickel silicate–nickel hydroxide composite with two types of morphologies for enhanced electrochemical storage performance. Mater. Chem. Frontiers **3**(10), 2090–2101 (2019)
124. D. Gu, X. Zhou, Z. Sun, Y. Jiang, Influence of Gadolinium-doping on the microstructures and phase transition characteristics of VO_2 thin films. J. Alloy. Compd. **705**, 64–69 (2017)
125. N.M. Ndiaye, T.M. Masikhwa, B.D. Ngom, M.J. Madito, K.O. Oyedotun, J.K. Dangbegnon, N. Manyala, Effect of growth time on solvothermal synthesis of vanadium dioxide for electrochemical supercapacitor application. Mater. Chem. Phys. **214**, 192–200 (2018)
126. I.N. Reddy, A. Sreedhar, J. Shim, J.S. Gwag, Multifunctional monoclinic VO_2 nanorod thin films for enhanced energy applications: photoelectrochemical water splitting and supercapacitor. J. Electroanal. Chem. **835**, 40–47 (2019)
127. Y. Zhang, X. Jing, Y. Cheng, T. Hu, M. Changgong, Controlled synthesis of 3D porous VO_2(B) hierarchical spheres with different interiors for energy storage. Inorg. Chem. Front. **5**(11), 2798–2810 (2018)
128. S. Chen, H. Yu, L. Chen, H. Jiang, C. Li, Defect-domains enabling VO_2 nanosheet arrays with fast charge transfer for 3.0 V aqueous supercapacitors. Chem. Eng. J. **423**, 130208 (2021)
129. Y. Dong, Z. Zhu, Y. Hu, G. He, Y. Sun, Q. Cheng, I.P. Parkin, H. Jiang, Supersaturated bridge-sulfur and vanadium co-doped M0S2 nanosheet arrays with enhanced sodium storage capability. Nano Res. **14**(1), 74–80 (2021)
130. Y. Zhang, L. Tao, C. Xie, D. Wang, Y. Zou, R. Chen, Y. Wang, C. Jia, S. Wang, Defect engineering on electrode materials for rechargeable batteries. Adv. Mater. **32**(7), 1905923 (2020)
131. Q. Xu, Y. Liu, H. Jiang, Y. Hu, H. Liu, C. Li, Unsaturated sulfur edge engineering of strongly coupled MoS_2 nanosheet-carbon macroporous hybrid catalyst for enhanced hydrogen generation. Adv. Energy Mater. **9**(2), 1802553 (2019)
132. Y. Xie, L. Zhou, C. Huang, H. Huang, J. Lu, Fabrication of nickel oxide-embedded Titania nanotube array for redox capacitance application. Electrochim. Acta **53**(10), 3643–3649 (2008)
133. C.C. Raj, R. Prasanth, Review—advent of TiO_2 nanotubes as supercapacitor electrode. J. Electrochem. Soc. **165**(9), E345 (2018)
134. K.M. Thulasi, S.T. Manikkoth, A. Paravannoor, S. Palantavida, M. Bhagiyalakshmi, B.K. Vijayan, Facile synthesis of TNT-VO_2(M) nanocomposites for high performance supercapacitors. J. Electroanal. Chem. **878**, 114644 (2020)

135. P. Gao, R.J. Koch, A.C. Ladonis, S.T. Misture, One-step solvothermal preparation of Mn-doped $VO_2(B)$ nanosheets for high-performance supercapacitors. J. Electrochem. Soc. **167**(16), 160523 (2020)

136. W.J. Zhang, M.X. Sun, X.H. Yan, M.Y. You, Y.L. Li, Y.H. Zhu, M.Y. Zhang, W. Zhu, M.S. Javed, J.M. Pan, S. Hussain, Pseudocapacitive brookite phase vanadium dioxide assembled on carbon fiber felts for flexible supercapacitor with outstanding electrochemical performance. J. Energy Storage **47**, 103593 (2022)

137. T.T. Nguyen, J. Balamurugan, V. Aravindan, N.H. Kim, J.H. Lee, Boosting the energy density of flexible solid-state supercapacitors via both ternary NiV_2Se_4 and $NiFe_2Se_4$ nanosheet arrays. Chem. Mater. **31**(12), 4490–4504 (2019)